ARTIFICIAL INTELLIGENCE AND PLAYABLE MEDIA

This book introduces readers to artificial intelligence (AI) through the lens of playable media and explores the impact of such software on everyday life.

From video games to robotic companions to digital twins, artificial intelligence drives large sectors of the culture industry where play, media and machine learning coexist. This book illustrates how playable media contribute to our sense of self, while also harnessing our data, tightening our bonds with computation and realigning play with the demands of network logic. Author Eric Freedman examines a number of popular media forms—from the Sony AIBO robotic dog, video game developer Naughty Dog's *Uncharted* and *The Last of Us* franchises, to Peloton's connected fitness equipment—to lay bare the computational processes that undergird playable media, and addresses the social, cultural, technological and economic forces that continue to shape user-centered experience and design. The case studies are drawn from a number of related research fields, including science and technology studies, media studies and software studies.

This book is ideal for media studies students, scholars and practitioners interested in understanding how applied artificial intelligence works in popular, public and visual culture.

Eric Freedman is Professor and Dean of the School of Media Arts at Columbia College Chicago. He is the author of *The Persistence of Code in Game Engine Culture* (2020) and *Transient Images: Personal Media in Public Frameworks* (2011). He serves on the editorial boards of the *International Journal of Creative Media Research* and the *Journal of Communication and Media Studies* and on the Advisory Board of the Communication and Media Studies Research Network.

ARTIFICIAL INTELLIGENCE AND PLAYABLE MEDIA

Eric Freedman

Routledge
Taylor & Francis Group

NEW YORK AND LONDON

Cover image: Yuliash/Getty Images

First published 2023
by Routledge
605 Third Avenue, New York, NY 10158

and by Routledge
4 Park Square, Milton Park, Abingdon, Oxon, OX14 4RN

Routledge is an imprint of the Taylor & Francis Group, an informa business

Library of Congress Cataloging-in-Publication Data
A catalog record for this book has been requested

ISBN: 978-1-032-12545-9 (hbk)
ISBN: 978-1-032-12481-0 (pbk)
ISBN: 978-1-003-22507-2 (ebk)

DOI: 10.4324/9781003225072

Typeset in Bembo
by Apex CoVantage, LLC

CONTENTS

FIGURES

ACKNOWLEDGMENTS

This book emerged from a very uneasy space, during a period that forced many of us into hypermediated states, divorced from others and experiencing the world through the lens of computer-mediated communications. To safely occupy that space is a point of privilege; for me, this meant having the security and the means to take the time to be immersed in new systems, new games and new work-related modes of connectivity, while studying their pattern languages. While I have been shielded from some of the harshest pandemic realities, I am mindful that the socioeconomic fallout of COVID-19 will ripple through society for decades and have a disproportionate impact on already vulnerable communities; and that mindfulness has forced me to consider the value of writing about artificial intelligence and playable media at a time when many scholars have turned to the more overtly pernicious forms of intelligence associated with data mining, deception, displacement and control. Figures as divergent as Stephen Hawking and Elon Musk have urged us to take artificial intelligence seriously as a threat to humanity, and have prophesized that an unexpected explosion of machine learning will pose an existential risk. In his radio lecture "Intelligent Machinery, A Heretical Theory," a broadcast that aired around 1951, British mathematician Alan Turing postulated that self-learning machines might take control of the world. And yet, the most common thought exercises for the brain without a body have been learning languages, performing complex mathematical equations and playing games (Turing, 1948). So, I remain hopeful about our augmented future, and about using artificial intelligence to solve the complex multi-system problems that threaten our very existence.

I thank my external reviewers for lending their critical eyes to my research and pushing me to clarify my arguments; their feedback strengthened the manuscript by making me a more thoughtful writer. I thank the anonymous gamers who have served as a surrogate community. Their collective humor and camaraderie continue to outweigh their worst behaviors. Many of us danced together in game space to celebrate our shared victories. My colleague and friend Heather Hendershot invited me to speak to graduate students in the MIT Comparative Media Studies Program, and that lively conversation pushed me to clarify several of my arguments. Of course, the strongest editorial voice came from my editor Sheni Kruger at Routledge, who expressed enthusiasm for this project at its outset and shepherded the book forward with an always patient and diligent Emma Sherriff at the press. Finally, I thank my husband, Ryan Ratliff, who has once again encouraged me to follow new ideas and has always given me the time and space to do so.

1

COMPUTATION MEETS PLAY

The History of Playful AI

This book pulls on a number of interrelated threads—play, media and artificial intelligence—and ties these together to survey the field of objects and experiences that we can group together as playable media: adaptive or variable systems that learn from and respond to player engagement by reading each interaction as a data point. By understanding their players as part of a pattern language, as imprints on an underlying codebase, playable media can create increasingly detailed informatic loops. Most research on playable media seems to privilege video games, but the field is more expansive and varied; beyond a set of fixed object lessons, we can use the term to explore the computational structures that undergird a broad range of responsive and adaptive media forms, and consider how those structures are used to establish a social contract. This book furthers the important task of examining the impact of software on everyday life as it traces the industrial development and migrations of artificial intelligence (AI) and the quality of our playable attachments with video games and other data-driven textual instruments that invite, structure and adapt to play (Wardrip-Fruin, 2005). AI provides the functional intelligence that makes computational systems more open, playful and connected. AI produces contextual awareness, socializes big data and creates an infrastructure for engagement; and while AI algorithms often operate like a black box, their outputs (their data mining activities) are often expressly visual. While the historical development of AI has been tied to making machines more intelligent, whether to aid or supersede human problem-solving, the curious intersections of artificial intelligence and playable media seem to offer up other value propositions, other generative algorithmic possibilities and other directives for media technologies. Indeed, our most common media technologies, those smart devices that seem to have the highest personal value and those systems that our society is built on, were developed from AI research. The central

DOI: 10.4324/9781003225072-1

goal of this book is to help media scholars, practitioners and students understand how AI works in relation to personal and collective agency and to see its influence on visual culture, while foregrounding the cultural knowledges and values that have been associated with public-facing deployments of AI. AI algorithms are designed by humans as repeatable computational constructs to be integrated into complex human-designed software systems and intelligent machines that provide the grounds for play. The concept of playable media, media that are created with and activated by a wide range of intelligence systems, is used throughout this manuscript to invoke a sense of freedom—a freedom that exists in those game and non-game environments that are relational rather than transactional. Although these operations are often intertwined, playable media are performative systems that use machine intelligence to make the world more dynamic and discoverable as they in turn discover their interactants. Playable media are bound to conceptually unified data sets and actions, but they are also part of a much larger network of algorithmic decision-making systems. Playable media stand as conditional supertexts, influenced by a series of interdependent computational intelligence strategies and narratives. Playable media are not walled gardens or closed platforms; they are connected to other systems. Although their data flows are commonly regulated to protect corporate capital, playable media draw from and contribute to a much larger data ecosystem, an ecosystem that has been transformed by advancements in machine learning and the deep learning of neural networks as well as significant increases in computing power and storage. Playable media are not immune to the broader socio-technical challenges of the data ecosystem—of its architecture, management and governance—and the distinct motivations of data producers and consumers (Gröger, 2021). Playable media are differentiated from other media in several ways. Play sets up one constitutive boundary that indicates a particular form of engagement with the media object (or experience) at hand; and the suggestion that the media form is playable establishes another boundary, and indicates that it has qualities that situate it as separate and apart from non-playable media. Playable media invite us to participate and, in many cases, require our participation to close a circuit of relations; they need our input, our data, to unlock their latent algorithmic potential. Artificial intelligence provides the spark that makes them more responsive and engaging; yet across playable media forms, we can observe distinctly different forms of AI. For example, the human-level intelligence agents we find in many video games are altogether absent in digital twins. Playable media embody a broad spectrum of AI agent-to-human interactions (Laird and van Lent, 2001), and while game AI is more attentive to what its users are doing, the distributed systems of digital twins are less focused on their users and more on the aggregated presentation of real-time data.

Existing scholarly literature on artificial intelligence proceeds from one of three perspectives: the computer sciences, the social sciences or design practices (for content developers). This book provides a bridge between these disciplines and the broader fields of media and technology studies. My approach does not assume any of the same knowledges (in computer science or social science) and is designed for readers who study and interact with media but do not routinely consider its operating systems; and this book is purposefully structured to expand outward from video game studies to consider other forms of playable media that are adjacent to and connected to video games as they are driven by common information architectures (game engines and AI subsystems). Video games provide an easy entry point for considering play as the product of an information architecture before proceeding onward to tease apart the interdependencies between programming and play; and it is logical to approach AI by way of video games, as they have been used to benchmark advances in computational intelligence and as arenas to play through more immersive and expressive experiences with machine intelligence (Fuchs and Sudmann, 2019). Moreover, the general push in engine-based game and media development, asset creation and interaction toward realism, and the exponential growth of the video game market has positioned video games as a highly visible showcase for state-of-the-art AI. As such, this book is ideally situated as a useful addition to the fields of media studies, video game studies and software studies, as I have pursued a path that connects the discussions of data and algorithmic systems to a series of everyday media objects as well as advanced media forms such as robotics. The manuscript broadens the field of AI beyond video games to other industries that draw from the core principles and technical underpinnings of interactive media, and is designed to help media students and scholars understand how computational systems have shaped playable culture and how they are developed as part of interrelated media, design and information systems; and while the focus of this work is on play, as most scholars of play agree, play is not without consequence. Andrew Goffey (2008) notes that algorithms act transversally through data on both humans and machines; although they are statements structured as machinic discourse, their effects are broad and real (Goffey, 2008). Algorithms perform transformative and generative operations on collocated data and, in turn, inform how we act. We may of course draw distinctions between AI-based media objects and those that are driven by simple algorithms; artificial intelligence implies a greater dependency on data within a machine-learning model that can continue to improve itself through training data. While an algorithm is an automated instruction, a sequence of statements or a series of complex mathematical equations, machine learning connects those statements to a structured data set to execute a number of fixed tasks, while artificial intelligence systems allow for the additional complexity of new situations and uncertain variables—the introduction of unstructured data that can be structured, assimilated and acted upon.

Artificial intelligence frameworks undergird large sectors of the culture industry that depend on systems of machine learning and logic, and where play and structure coexist; AI frameworks are used as building tools to guide and automate the design process, and as behavioral tools to guide the experience of play. AI is deployed as an open-source tool but also as an element of proprietary black box intelligence systems that obscure their operations and motivations. Through a series of case studies that range from video game development to connected home fitness machines, to other institutionally specific systems and prototypes, the goal here is to introduce readers to AI through the lens of playable media and within a range of media platforms, where decision-making is informed by algorithms, and to illustrate why AI matters, as the technical patterns and structures of machine learning migrate to other industries that foster and learn from consumer interaction. AI is fundamentally a computational practice, focused on efficiency and rational problem-solving, but it is also a representational practice and a social practice. AI is central to creating systems and architectures that complement authorial control with the generative capabilities and playful possibilities of autonomous responsive player interaction (Bogost et al., 2005). AI effectively broadens what we commonly understand as meaningful play or the affirmative values attached to player choice and action within a discernable ecosystem (Salen and Zimmerman, 2004).

A player's activity, however purposeful within a responsive system, can be socialized and mined, and can be used as a communicative act with far-reaching consequences. *Voilà AI Artist*, a photo editor app that became fairly popular on social media platforms in the United States in mid-2021, draws from several of the core components of artificial intelligence, parsing, assimilating and modeling big (visual) data into a number of discrete styles through well-proven deep-learning techniques. The iOS- and Android-compatible app transforms personal headshots into 2D and 3D cartoons, caricatures and painterly portraits. The privacy practices of developer Wemagine.AI allow the company to use customer data to improve its services (a policy statement that is decidedly ambiguous) and push target-based advertising from third-party partners. *Voilà AI Artist* produces two distinct outcomes that are driven by a common data set—one that (re)faces the player and one that supports the developer's platform-based business model. Moreover, the application encourages us to play with facial mapping in an economy increasingly reliant on facial recognition technologies. Playable media leave long data trails. They are meaningful localized experiences that happen within discernable systems of exchange that carry broader cultural import as they harvest and repurpose player data; playable media are both relational (for the player) and transactional (for the developer, publisher or service provider). Playable media are data-driven commodities that regularly invite us to act as technobiographic subjects (Freedman, 2011), engaged with and inscribed by information technologies. There is an art to data governance; technosocial relations are built through systems of convenience that establish a familiar disposition, in this case, to artificial intelligence, through "multiform tactics" (Foucault, 1991, 95) that bend us toward platform dependence and an economy undergirded by computational analysis. When playable media are embedded in existing platforms and devices, they carry the ideological weight of those infrastructures and technologies, and they are functionally contoured to meet a series of pre-existing hardware and software specifications and a number of domain-specific data practices; the ongoing calls for the federal regulation of social media are largely calls for algorithmic transparency, and should alert us to the uncertain terms of play in such spaces.

Artificial intelligence is a broad field, encompassing an ever-widening array of perspectives, technologies and applications, and is embedded in both networks and objects. AI has been attached to existing cultural anxieties about the inscrutable power of technology and has in fact furthered existing social inequities, especially as a tool in the labor force and in data-driven governance, and those same concerns follow the various migrations of artificial intelligence into playable media. AI technologies are fundamental to the architectures of platform capitalism and to those businesses that provide the hardware, software and data foundations for others to realize and maximize their operations, refine their management techniques, further their efforts with automation and, in turn, undermine the agency of their workers. Artificial intelligence shapes the flow of information and influences decision-making. The development of new algorithms represented by the field of cognitive computing continues to further the cross-industrial applications of AI. Cognitive computing pushes AI in a more specific direction to extract the relationships from data, understand and learn from these patterns, and drive us toward significant solutions; cognitive computing is shaping our understanding of the objective world. Artificial intelligence is an expansive term, and the turn toward cognition focuses our attention on the interpretive dimensions of data-laden processes. But for the purposes of this manuscript, cognition is understood as a facet of many (though not every) AI frameworks. AI undergirds recommendation engines, machine-learning systems, robotics and the Internet of Things (IoT). AI must be understood in the context of (unstructured) big data and the demands of a digital economy accelerated by cloud computing and the automation and information requirements of a functional and responsive IoT. Data must be turned into knowledge and action, and artificial intelligence makes this possible in both data and decision science; the former applies a statistical lens to data, looking for patterns and finding causal relationships, bringing order to and learning from big data, while the latter foregrounds the stakeholder's question and frames data as part of the decision-making process to solve a localized problem. This distinction is one of emphasis of whether the business problem comes first and is followed by the algorithmic pursuit. The congruence between data and decision science, the connective thread between building a data framework and taking calculative steps based on data analysis, and between pattern recognition and practice lies in the behavioral or cognitive attitude toward the work, where data find a cultural, industrial or civic purpose.

The focus here is the impact of AI on playable media, with the latter defined by computationally driven media systems that are dependent on both human interaction and machine learning (which we can parse as two distinct forms of agency). AI is a product of human intelligence and as such aligned with our desire to make sense of (and control) a complex world. The focus on AI is not a call to study the technical infrastructures of playable media and abandon their formal devices, visual tropes and narrative pleasures; we should always try to pay equal attention to computation, representation and interaction, and understand the interdependent layers of digital media. Playable media are environmental systems, populated with agents that induce us to engage with them. This manuscript starts with the field of video game software development, as video games have served as testing grounds for artificial intelligence, and then proceeds to build a more expansive view of software architectures that foster and learn from playful interaction—software embedded in more common consumer objects and as a driver of deeper brand engagement and deeper dependencies with responsive media. Playable media are defined here as computational and performative systems and experiences, developed from a number of machine layers (data sets, programs, assets, interfaces and objects) that invite and structure play and produce meaning partly through their mechanics and their computational flexibility; the informatic layer of playable media is enveloped by accessible and communicative context-dependent design. As I have established the focus here on playable media, I must also acknowledge that, in doing so, I am setting one of several possible frames for understanding and assessing artificial intelligence, although my intent is not to set a limiting horizon. I am highlighting one paradigm, performing one intervention that might help us reread algorithms more generally and understand their cultural weight, but this focus on playable media aligns with the general interdependencies of AI systems. AI systems require an integrated hardware and software foundation for writing and training machine-learning algorithms, and they work by ingesting and analyzing large amounts of data to find correlations and patterns that can generate or predict future states.

My approach to computational media is humanistic, an effort to understand how individuals act with and are acted upon by playable media, although throughout this manuscript, I also consider how integrated human-centered design strategies have expanded the technical infrastructures of game development, allowing these existing code frameworks to flow into other media forms. The strategies for computational representation that drive games and other playable media forms are grounded in what Michael Mateas and Noah Wardrip-Fruin (2009) refer to as "operational logics"—the abstract operations that underwrite gameplay and serve as guideposts for engagement. These operational logics are often knowable and observable as they tie algorithmic work to textual work. Playable media include game and non-game objects and a body of performance-based experiences driven by computation; as Wardrip-Fruin (2005) notes, it may be more productive to consider how an object is played rather than attempt to locate it within more formal and prescriptive game definitions. This is a study of intermedial relations, a study of the set of computational characteristics that exist across playable media, and a study of the fundamental transformation of playable media and play more broadly by artificial intelligence.

Understanding that video game software is a particular type of code object that has been transformed by AI, we can begin to locate similar industrial transformations forged by parallel AI frameworks. The general turn from computer and video games to broader forms of playable media within this manuscript is designed to highlight how we play, and to understand the conditions of play across a number of manufactured code objects. From this perspective, we can understand gameplay as a continuum of experience that productively expands the field of games without invoking the overused concept of gamification. This manuscript does not draw from existing video game ontologies, although it examines video game software practice; rather, it is focused on the persistent code work of play. I have suggested that playable media are performative to highlight the code work necessary to produce interaction but also to advance the thesis that the play functions of playable media are not simply grounded in their interactivity but also how they advance selfhood by solidifying a number of behavioral strategies that produce knowledge, mastery and control, and call out both our social selves and our embodied selves. Thomas Henricks (2014) suggests that "play is an exploration of powers and predicaments" (204), of testing ourselves through a broad range of material environments. AI generates the patterns and processes that challenge us in playable media.

The book is organized as a series of case studies or object lessons that make clear the pervasive use of AI across a broad range of industries that have driven not only the colonization of play but also its permutations; through each study, the book teases out several of the primary subfields of AI (machine learning, neural networks, speech recognition, natural language processing and computer vision) that are driving each industry, and locates these in distinctly integrated hardware and software solutions. By linking personal and dynamic data (reading the state of the user through the conditions of interaction), computational systems are consistently used to structure and extend play, deepen user engagement and shape social space. But this studied attention to situated consumer practices has led us away from playful inefficiency. Machine learning by its very nature is directed toward realizing more efficient systems, and while play is commonly oriented toward exploration and discovery, artificial intelligence has realigned play with the demands of networkable logic.

Code continues to create new horizons for media studies. We are living in a machine-readable world that can be acted on by software independent of human control and in which datafication increasingly turns questions of self over to new products and systems of networkable data—products and systems that shape visual culture, thought and action (Freedman, 2020). This manuscript centers artificial intelligence as a pathway for media studies scholars and practitioners to navigate the broad terrain of software practice, and is oriented as a media-centric approach to the landscape of artificial intelligence. The case studies presented here lay out a methodology for situating machine learning as an embedded practice and as part of a broader lesson about connecting algorithms to material outcomes and about modeling behavior. The particular material outcomes considered here fall under the purview of playable media and follow a similar design practice—remapping a particular set of code frameworks, those commonly associated with video games, onto other media forms to produce computationally flexible and responsive user experiences. These experiences are saturated with pre-existing game logics and have benefited from the research driven by many of the technical demands of video games associated with persistence and real-time feedback, with building and navigating complex worlds, and with realizing sociability and networkability.

Playing With Intelligence

To properly situate playable media, we need to focus on both the science and the matter of play (the technical ground, the content and the experience). The video game industry is not ground zero for artificial intelligence; there are much earlier intersections between computation and media culture, and the incremental application of artificial intelligence to playable media has been happening since the birth of the field in the mid-1950s, when it was largely the domain of the computer sciences. Alan Turing (1950) invoked the concept of play with the "imitation game," as he pursued the problem of machine intelligence and imagined a computational future that would bring into question our stated assumptions about intelligence. To approach the question of whether machines can think, Turing devised a game played with three people, notably gendered a man, a woman and an interrogator of either sex (Turing, 1950)—a game whose goal is also attached to its gendered axis, as the interrogator aims to determine the identities of the other two participants. Moreover, to address the primary research question, Turing asks what will happen when the man is substituted by a machine. Significantly, Turing's thought experiment locates the problem of machine intelligence not in the mechanics of computing but in the development of suitable programming that can benefit from the principles of machine learning. Turing's "Intelligent Machinery" (1948), a report for the National Physical Laboratory in London, identifies the importance of learning from context and of exploration. Outlining the task of building a robust "thinking machine," Turing proposes two methods, the first of which is a mobile object-oriented construct: "In order that the machine should have a chance of finding things out for itself it should be allowed to roam the countryside" (Turing, 1948, 9). While the second method is to build a brain without a body, both propositions emphasize the importance of education, of learned behavior over time and of what can be more broadly understood as developing cultural intelligence. In the *Atlas of AI*, a critical examination of the political power of artificial intelligence, Kate Crawford (2021) draws out the particularities of how, precisely, intelligence is made and the dynamic work performed by the constellation of institutions, interests (financial, cultural and scientific) and networks that construct it. Crawford suggests that intelligence is built and embodied through a series of wide-ranging relational ecologies (Crawford,

2021); to deconstruct AI, grasp its operations and unpack its labor and its power, we need to understand these relations and account for the unwavering influence of social, cultural, historical and political forces on information technologies as they are developed and put to use.

AI research has advanced through a number of experiments with playable culture, all of which have tested the limits of machine learning through games and which firmly ground AI in the networked communications of self-learning machines: the 1997 six-game chess match won by IBM's Deep Blue computer against Russian player Garry Kasparov, the 2011 Jeopardy tournament won by Watson (another IBM supercomputer), 2013 DeepMind research with deep learning, computer vision and pixel recognition and the Atari 2600 (studies that were expanded at Google after the company acquired DeepMind in 2014), the 2016 Go matches won by Google's AlphaGo program against South Korean player Lee Se-dol, 2017 Microsoft (Maluuba) playtesting research with multi-agent reinforcement learning and *Ms. Pac-Man*. More recently, video games have been a testing ground for furthering the work of generative adversarial networks (GANs). *GAN Theft Auto* is one such effort, an experiment that uses the AI tool GameGAN to recreate a highway environment pulled from *Grand Theft Auto Five* (an open-world game released by Rockstar Games in 2013) and produce a short playable demo. GameGAN is a generative adversarial network created by NVIDIA that can learn to visibly imitate a specified video game by ingesting screenplay and keyboard actions during the training process. By training the generator on a sample data set (in this case, a discrete gameplay environment), the tool can produce parallel interactive content based on fairly complex source material. The auto-generative abilities of artificial intelligence have been similarly furthered by a number of non-game experiments. In 2018, DeepMind researchers trained their neural network to imagine a three-dimensional scene by learning from a series of two-dimensional images taken from different viewpoints; the generative query network (GQN) proved its ability to render details from static images as it mapped their spatial relationships, successfully predicting what objects would look like from other perspectives while teaching itself about color, lighting and texture. Whereas supervised learning requires large data sets that are manually annotated so that a computer can learn how to interpret new images, a process that does not leave room for more associative observation, interaction and contextual insight, the GQN fills in gaps in the data that might result from occlusion or limited scene perspectives without any explicit human supervision: "The GQN takes as input images of a scene taken from different viewpoints, constructs an internal representation, and uses this representation to predict the appearance of that scene from previously unobserved viewpoints" (Elsami et al., 2018, 1204). The GQN effectively stitches together three-dimensional representations and situates objects in space, but it is doing more than stitching together a series of images—it is, in effect, imagining what it has not seen.

There is a large volume of applied research on image-based rendering systems that can perform similar unstructured view synthesis with or without deep learning—systems that can map, blend and construct new scene views (Riegler and Koltun, 2020). Collectively, these systems are rendering visual culture from a perspective that, as it advances, is less and less determined by human intervention, although it was originally conceived and developed from within an anthropocentric paradigm and prone to human and humanistic biases. By extension, AI-driven systems of representation are transforming screen culture, impacting the visual grammar, organizational labor, technical structures and industrial models of film, television, video games and related media. The machine learning of GANs requires training two opposing but complementary algorithms; a generator learns to create plausible data, while a discriminator learns to discriminate between the generator's fake data and real data. The two algorithms work in tandem to create a system that gets better over time at generating realistic images, voices and videos. The goals of intelligence research are commonly understood as twofold: to simulate human intelligence to such a degree that machines might replace human beings and occupy their locations within an organization, a representational framework or a development pipeline, and to yield discrete programs that might augment human intelligence and assist their end users in carrying out their assigned tasks, providing solutions to practical problems (Mirowski, 2003).

In 2018, a team of NVIDIA researchers proposed StyleGAN, a style-based generator architecture tasked with analyzing and synthesizing existing facial image data to produce novel photo-realistic images (Karras et al., 2019). In 2019, software engineer Phillip Wang used NVIDIA's open-source generator to create *This Person Does Not Exist*, a website that conjures its fake portraits from existing image data, and produces new human faces with each page refresh. This data-driven image modeling is based on a degree of normalization that is achieved by training an algorithm on existing data sets and outputting in ways that match stylistic conventions and expectations; the goals of the system are an unsupervised image-to-image translation and the production of a novel image that conforms to common facial shapes and features while invalidating nonconforming edge cases. The StyleGAN operates within and produces acceptable structures, and although it is a fascinating visual tool, it carries with it much starker undercurrents. GANs are often at the root of deepfakes, as they couple machine learning to large data sets to try to replicate real-world image patterns until the fakes are indistinguishable from the originals. GANs can develop unique human faces that pass for real people. A series of very real anxieties have been associated with the hidden operations and imperatives of AI-driven systems.

Artificial intelligence tools are deployed within institutions that are historically marked by systemic discrimination—housing, the workplace, the criminal justice system and our financial institutions—and bias is often baked into the outcomes of what AI is asked to predict. The data used to train machine intelligence are commonly under-representative of people of color, women and other marginalized groups, and those fault lines impact the design, development, implementation and outcomes of AI across a broad range of tech and tech-dependent industries. For example, lending algorithms, which rely on hidden proxy variables for credit worthiness to create a composite sketch of an individual, have led to discriminatory banking practices in the home mortgage market. As decision-making moves from humans to machines, it becomes more difficult to extract the signaling processes embedded in automated systems that are purposefully designed to create more efficient workflows (Benjamin, 2019). Even open-source algorithms have limited transparency if their training data remain hidden.

To illustrate the operations of a number of AI systems, including chatbots, and to underscore several of the structural issues of these systems, such as bias in data modeling, IBM's "Learn + Play" initiative invites players to participate in a series of simple games. IBM's "Machine Learning for Kids" is a hands-on activity kit that introduces children and teens to machine-learning concepts through a number of interactive data projects. "AI for Oceans," produced by Code.org, invites young learners to train an intelligent agent in pattern recognition to differentiate between marine life and manufactured waste products. These are just three of the many offerings in the company's "AI for Kids Catalog" (IBM, 2021), a series of lessons that incorporate play to illustrate the fundamental operations of artificial intelligence and machine learning and the practical applications of algorithmic systems. By providing these playable resources, IBM is helping young learners discover the social impact of algorithms, familiarizing them with a series of interrelated information processes that impact the world at large. Through these activities, IBM continues to extend the reach of its own research as the company builds out its intelligence portfolio.

There is a complex relationship between AI and its technologies, as self-learning does not reside in one set of objects. Computational systems exceed the limits of objecthood, which is why I open this manuscript by situating them in the field of play rather than in video games *per se*. Nevertheless, the AI research performed through computer games has been fundamental to expanding the field. This particular strand of AI research is designed to test algorithmic systems and to find new ways of teaching machine agents to navigate complex data-centric environments and work toward common goals. Computer games are a playground for testing, developing, modeling and performing collective intelligence, and this thread of AI research can accelerate systems work in other fields: agriculture, healthcare, manufacturing, neuroscience, robotics, transportation, energy and the environment. In turn, modern AI methods have significantly altered the landscape of video games by opening up new design practices and new forms of interaction.

The playful side of AI is perhaps most apparent in the "Internet of Toys"— a field dotted with intelligent animatronic devices that include Sony's AIBO entertainment robot. The first three generations of AIBO robotic dogs were designed and manufactured between 1999 and 2006; across that span, the AIBO moved through a number of model designs that were all capable of seeing the environment and recognizing spoken commands. In late 2018, Sony resurrected the AIBO concept with a network-enabled dog that features scaled-up machine learning. The three-year AI Cloud Plan that pairs with the product enables AIBO to upload its day-to-day experiences to Sony's cloud-based AI engine. The AI system analyzes and interprets each data draw before returning more evolved behavior patterns back to the AIBO to perform. The My AIBO application extends the feedback network to an interactive mobile interface that allows owners to manage, customize and even view a live stream of what the AIBO is seeing. Sony engineered the robot to "mature" (develop its autonomous behavioral systems) over the course of three years, to mimic the developmental learning patterns of a real dog. While the AIBO does not fulfill Turing's dream of a listening, seeing, learning and thinking machine that might roam the countryside, we can see this premise as a ground for the purposeful commodification of AI and the development of a network of operations that make possible computationally driven play. The AIBO is both a material architecture and an industrial system of artificial intelligence. But as a playable form, its experienced value exceeds its object status; in 2018, over 100 AIBO dogs were given a funeral in Japan's historic Kofukuji Temple, each tagged with its name, a description of its family owners and its place of origin. Although designed as an adaptive learning machine and an active communication agent capable of expressing its own temperament and emotions, the first generation AIBO Entertainment Robot (ERS-110) and its successors were purposefully designed as malleable maturational systems (stepping through babyhood, childhood, adolescence and adulthood) with open-ended futures, born to live with people, learn from the environment and become beloved family members. This robotic system realizes the full circuit of applied artificial intelligence—sensing, analyzing data and producing mechatronic outputs informed by the real world—and furthers the functional integration of AI into everyday life through its social bearings. The AIBO is part of Sony's grander entertainment ecosystem, designed to forge emotional connections between robots and humans and more broadly between humans and technology. Sony promotes the AIBO as "technology for being loved" (Sony, 2021) and artificial intelligence provides the computational axis for building such a bond and for furthering the work of embedded intelligence frameworks. The AIBO is a comfortable ancillary to Sony's other AI projects, including its Neural Network Console, an integrated software development environment (publicly released in 2017) for designing deep-learning neural networks, but the AIBO's AI insights are driven by data collection and analysis across a network, and this dependency has run up against public policy. The State of Illinois Biometric Information Privacy Act (BIPA) has prevented sales of the AIBO to Illinois residents. The AIBO collects facial recognition data through its cameras (the camera above its tail helps it map the owner's home, while the camera on its nose can remember and recognize up to 100 individual faces), and this biometric information places specific obligations on Sony under Illinois law. While not an assault on artificial intelligence *per se*, these regulations reveal the conflicts that emerge from the internal interdependencies of data-driven mechanized systems and point to the societal tethers of playable media. The Illinois state measure has tripped up Facebook in recent years. A class action lawsuit that was first filed in 2015, alleging that Facebook's practice of tagging people in photos using facial recognition without their consent violated Illinois privacy law, was settled in February 2021 with the tech giant being ordered to pay out $650 million, with approximately $345 million disbursed to Illinois class members. Similar lawsuits were filed on behalf of two Illinois residents in 2020 against Amazon, Microsoft, Google and IBM for using facial recognition systems and extracting and disseminating biometric data from personal photographs (to IBM's Diversity in Faces Dataset) without explicit consent.

Spot, an agile mobile robot with a dog-like body, became a global cultural phenomenon shortly after the Hyundai Motor Group's acquisition of a controlling interest in robotics designer Boston Dynamics. In June 2021, Hyundai released a whimsical promotional vignette welcoming Boston Dynamics to the family; the commercial video features Spot dancing alongside the bipedal robot Atlas and the Korean pop group BTS. Spot uses what Boston Dynamics refers to as athletic intelligence to independently traverse complex terrain, and this emphasis on autonomous mobility aligns the newly acquired product line with

FIGURE 1.1   Chief Priest Bungen Oi offers a communal funeral prayer for expired Sony AIBOs at the Kofukuji temple in Isumi, Chiba, Japan, in 2018. April 26, 2018.

Source: Nicolas Datiche/AFP via Getty Images

Hyundai's promise to build things that bring us closer together, a promise that in turn reflects the company's grander goal of realizing "progress for humanity" (Hyundai, 2021). These repeated anthropomorphic frames and these literal references to humanity, used in corporate mission statements and realized through a myriad of algorithmically adaptive bodies, allow us to comfortably live and play with machine intelligence, even as these machines are built to serve multiple and disparate industrial vectors—entertainment, manufacturing and the military. The same robotic dogs that dance to the beats of BTS have been used to defuse hazardous materials and survey potentially dangerous sites and situations. The Massachusetts State Police had Spot on loan from Boston Dynamics from August through November 2019 as a mobile remote observation device to assist in location-based surveillance and diagnostics; the New York Police Department began using Spot in December 2020, after painting the robot blue and renaming it Digidog, but terminated their lease after public outcries over its use as part of more aggressive policing in minority communities; in 2020, the Honolulu Police Department spent $150,000 in federal CARES (Coronavirus Aid, Relief and Economic Security) Act funds to purchase their Spot to perform touchless field screenings of homeless individuals at a government-run shelter, using Spot to run rapid temperature checks and provide expanded touch-free telehealth services; and in 2021, the Dutch National Police introduced Spot to their force to take the lead in drug lab investigations. These four applications of a common autonomous technology are bound by their attachments to a state apparatus, but each program is differentiated by its context. In Honolulu, Spot was acquired at a critical moment during the COVID-19 pandemic to keep an existing transitional shelter program running. Spot enabled the local police department to mitigate the spread of COVID-19 and meet a range of exposure challenges that were hampering the efforts of police officers, social workers and health care staff. Spot is more than an extension of governmentality; the mobile robot has also been used at a number of commercial sites to perform basic inspection tasks, carry payloads and collect data. What the popular promotional stunts belie and what the state-sponsored and commercial deployments of Spot convey are the robot's fundamentally program-based nature and the context-dependent values that are attached to its programming. Spot is not inherently a bad actor, but rather it is a freely programmed object inscribed by social agency.

In a second promotional video, "Spot's On It," released shortly after Hyundai's "Welcome to the Family" campaign, Boston Dynamics shared a longer cut of seven Spot robots once again dancing to BTS's "*IONIQ: I'm On It*" while performing a complex series of synchronized moves—a feat made possible through programming (with a Python-dependent Choreographer SDK) rather than functional obstacle avoidance. This is performance by design realized through calculated scripting for a choreographed sound stage shoot, which is quite distinct from performance in the field directed outward toward a graphed environment. In both videos, the robots appear to listen and respond to music; in actuality,

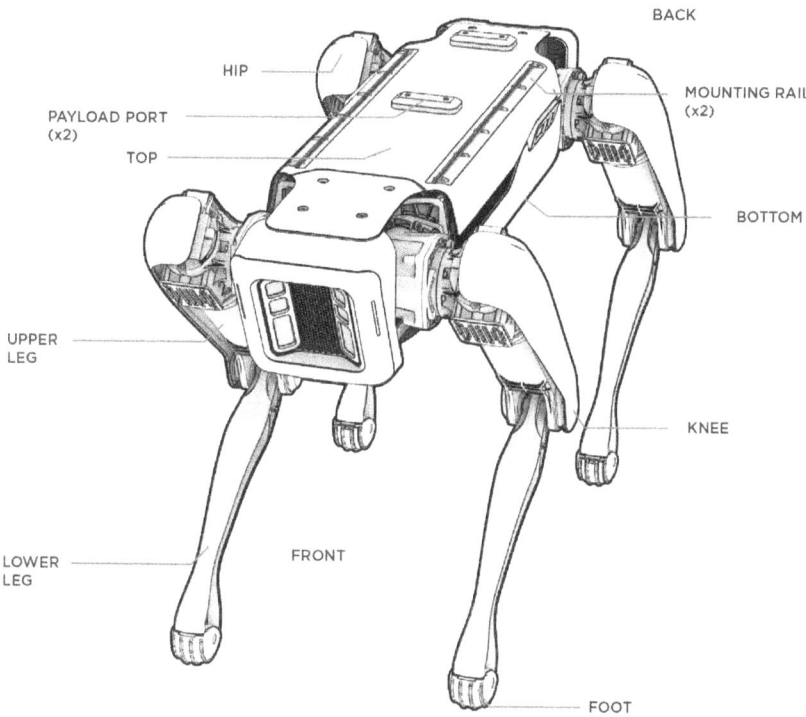

BACK

HIP

MOUNTING RAIL
(x2)

PAYLOAD PORT
(x2)

TOP

BOTTOM

UPPER
LEG

KNEE

LOWER
LEG

FRONT

FOOT

FIGURE 1.2   Documentation for Boston Dynamics Spot.

Source: https://dev.bostondynamics.com/docs/concepts/about_spot

Copyright 2021. Boston Dynamics

they do not, though they harmonize with each other and their articulations seem expressive; and while the dance also functions as an instance of accelerated life-style testing, moving seven robots (BTS is also a seven-member group) through an expanded movement scenario to test their durability, Spot does not understand what it is doing. In its commercial application, the world is a series of way-points, and the robot must be manually driven in order to map and record the environment before it can autonomously traverse it, but without AI, it does not truly learn from these graphs. While Spot was first released in 2019 without true onboard intelligence (beyond its autonomous mobility and pathfinding abilities), it was equipped with visual AI in 2020 through a partnership with Vinsa, a South Florida-based AI firm, following the company's participation in the Spot early adopter program.

Spot is part of a larger industrial ecosystem invested in robotics, AI and human–machine collaboration. The phrase "IONIQ" is itself part of a complex corporate economy, attached to the Hyundai electric car sub-brand announced in August 2020, fastened to a BTS and Hyundai cross-promotional music video released in September 2020 and rooted in earlier Project IONIQ Lab research—a Hyundai Motor Company project focused on advancing new mobility solutions by tracking a set of "megatrends" likely to influence the automotive industry. As part of the Lab's foundational white paper, Hyundai researchers envisioned a "hyper-connected society" with "context-aware individualization" made possible by big data and artificial intelligence (Hyundai, 2016). Spot enters this ecosystem as an open container of sorts, with the prospect of a Spot CORE AI on the horizon that will open up the possibility of machine learning, while third-party developers such as Vinsa have already added computer vision to the platform to extract actionable insight from visual data. Génération Robots, a robotics distributor with headquarters in France, suggests that these expansions "will transform your docile dog into a knowledge-hungry robot" (2021).

Playable media are malleable and expandable by design; they are not fixed hermetic systems (although they are often legally and functionally bound to a number of proprietary measures). Even with its commercial orientation, Spot is an open programmable system and a potential conduit for play. Likewise, most boy bands, including BTS, function as open programmable systems that are designed with an economic calculus, refined through various algorithmic processes, including manual pitch correction and Auto-Tune, and married to other assets within their corporate ecosystems. Artificial intelligence is not simply lodged in a series of Hyundai products; it is also a process and a pipeline. It is a pillar of factory labor and manufacturing, of speech recognition, natural language understanding and dynamic context-driven interaction, and of a series of autonomous technologies and lifestyle systems that include public transport. Electric vehicles, mobile robots and K-pop artists seem like strange bedfellows, pulled together as signs of the cultural relevance and influence of a South Korean conglomerate, but as surrogates and "spokes-objects" for a corporate value proposition, of a national brand image attached to human progress, they speak to our readiness to adopt strong emotional attachments to machines, if only to use them as extensions of our bodies. This is a relational tendency that Joseph Weizenbaum (1976) noted in the communal conversations with and about his ELIZA program in the 1960s, and expressed in some of the early enthusiasm among practicing clinicians about the progressive and democratizing possibilities of computer administered psychotherapy. Weizenbaum questions more broadly the inevitable impact of computers on society to consider the interplay of science and morality, a matter that complicates the simpler (yet still relatively complex) operation of translating life into formalizable systems (12). This complication speaks to what Herbert Simon and Allen Newell (1958) reference as the tension between heuristic and algorithmic processes and methods of problem-solving, and their prediction that digital computers might perform certain heuristic problem-solving tasks if they are able to learn and improve their performance on the basis of experience (7). Artificial intelligence research grew out of management science (to inform earlier operations research), and while it has expanded into studies of cognition, it remains attached to the development of expert systems that can emulate human decision-making in a specific domain (Simon, 1991). As a tool for developing, managing and running complex systems, AI has spread throughout society in ways that detach it from its more obvious relationship to numerical analysis and connect it to procedural goal-oriented activities (Simon, 1991, 129).

A Brief History of Artificial Intelligence

Sony and Boston Dynamics have produced two vastly different canine robots. The AIBO is a machine-learning system that encapsulates deep learning, one of two distinct fields of AI research—the other being symbolic AI. Both deep-learning and symbolic AI have been connected to unique forms of play. Symbolic AI was the primary research path from the 1950s to the 1990s, only to be eclipsed by deep learning. Symbolic approaches rely on preprogrammed knowledge and rules (they depend on and are limited by human inputs) that function as the building blocks of expert systems that can solve rudimentary problems (like chess) through heuristic search techniques. Deep-learning or machine-learning systems, such as the one that drives the AIBO, learn from data and use feedback loops to monitor and improve their performance. Machine-learning systems continue to learn as their algorithms establish correlations between inputs and outputs, and they benefit from neural network architectures, whereas symbolic reasoning systems are static programs built from a finite set of hard-coded relationships.

The rise of networked culture that followed from the expansion of the Internet in the 1990s fueled an AI powered by deep learning and neural networks that can leverage big data sets to solve problems of far greater complexity than a chess match; this is the matter of facial recognition, product and content recommendation engines, self-driving cars and a range of neural network-based systems that govern and expedite everyday life, systems that perpetually learn as their algorithms actively map inputs and outputs. AI researchers have, for the most part, abandoned the toys and puzzles that dominated the field from the 1950s to the 1990s, and turned their attention toward simplified rule-based simulations—the matter of video games. Video games provide a constrained space; they simplify the complexity of the real world, draw on recognizable patterns and are governed by software engines that can run and test advanced learning algorithms within a closed development loop. There are ongoing and shifting interdependencies between the fields of AI research and play, as AI has been used to meet the demands of certain design objectives and to satisfy the conditions of play.

The term "artificial intelligence" can be traced to a 1955 proposal written by John McCarthy, Marvin Minsky, Nathaniel Rochester and Claude E. Shannon for "The Dartmouth Summer Research Project on Artificial Intelligence," a proposal that led to the first academic conference on the subject held the following year. But work on artificial intelligence and the question of whether machines can truly think can be traced back even earlier to Vannevar Bush's (1945) seminal work "As

We May Think" and Alan Turing's (1950) paper on "Computing Machinery and Intelligence." These represent efforts to use machines to amplify human knowledge as well as refine their own behaviors, to learn to do intelligent things and to have the capacity to mimic human knowledge processes. The history of AI is, of course, intertwined with the history of computing, and it is generally agreed that the modern digital computer has its roots in World War II with the Bletchley Park Colossus machines that were built to mechanize the deciphering of coded transmissions. Parallel to the wartime work at Bletchley Park, the ENIAC machine was developed within the United States Army Ordinance Department. Computation was critical not simply for decoded communications but also for predicting the courses of aircrafts and the trajectories of missiles (Whitby, 1986, 117). While the computer has not remained an exclusively militarized apparatus, this history helps us understand how and why computers must be examined as cultural artifacts, as large commercial organizations and state institutions have evolved in relation to the development of efficient information-gathering networks driven by obedient programs. The goal-oriented nature of computational algorithms are, in part, a product of the historical ties and symbiotic attachments between the military and the computing industry and the complementary execution of operations that depend on and can produce fast and accurate information handling (Whitby, 1986, 117).

The summer 1956 conference at Dartmouth brought together a number of noted scholars and proved critical to furthering several strands of early research on computer reasoning and the importance of more openly conversational computer languages. McCarthy, Minsky, Rochester, Shannon and Turing were joined by Allen Newell and Herbert Simon. Newell and Simon were working with John Clifford Shaw at the Research and Development (RAND) Corporation in the mid-1950s, where they combined their talents in computer science, cognitive psychology, economics, political science and systems programming. Together they developed both the Information Processing Language (IPL), one of the first AI programming languages, and the Logic Theory Machine (LT), a cognitive simulation program written in IPL to run on JOHNNIAC, one of RAND's high-speed digital computers. Newell and Simon demonstrated the LT at the Dartmouth session, and while their presentation garnered little critical response, their work had a significant impact on the developing field of information processing. Newell, Simon and Shaw left a legacy at RAND, producing a number of the key building blocks for AI-related research (several of which were funded by the U.S. Air Force) including a chess playing program and the General Problem Solver (GPS). Collectively, their work pulled together a number of distinct research areas: symbolic processing, heuristic search, problem-solving, learning and cognition (Ware, 2008). Not incidentally, this work happened in tandem with RAND's involvement in designing and building digital computers and developing advanced programming techniques. Throughout the 1960s, RAND provided a center for natural-language research as well as the creation of human–machine interfaces, supporting the two primary strands of scientific inquiry that have driven the field of interactive computing and produced the computational environments that are central for designing and testing AI systems.

John McCarthy, a computer scientist at the Massachusetts Institute of Technology (MIT), adapted existing work on IPL to build LISP (List Processor) in 1958, which has become one of the standard languages for AI development in the United States. IPL and LISP are complemented by other languages, including Prolog (Programming in Logic), developed in 1972 by a French research team led by Alain Colmerauer and Philippe Roussel. Prolog was born from a project focused on processing natural languages, whereas IPL and LISP arose from more traditional investments in the computer sciences and the matter of programming. Yet the researchers that developed IPL, LISP and Prolog all held a common interest in bringing natural language processing to more direct human–machine communications. McCarthy (1959) succinctly captured the early challenges associated with building intelligent machines when he suggested: "In order for a program to be capable of learning something it must first be capable of being told it" (79). AI has evolved largely because of the availability of complex programming languages that expand on the native operations of machines. In 1959, IBM researcher Arthur Samuel developed a self-learning checkers program on an IBM 701 computer using its native machine language, but machine learning has evolved significantly with the introduction of new programming languages, including Python, a high-level general-purpose language first released in 1991. The development of AI is grounded in solving the language problem, iterating higher-level languages that escape the simpler and more restrictive operations of machine languages and begin to approximate more natural and expressive systems. IPL was the first language to be developed for the purpose of creating AI applications, only to be replaced by LISP, a higher-level and more scalable language. With the language problem largely addressed, AI research advanced through the discovery of more universal applications—research associated with the development of expert systems in the 1980s.

In his opening remarks at the first international symposium on artificial intelligence, a 1958 gathering on "The Mechanization of Thought Processes" held at the U.K. National Physical Laboratory, physicist Gordon Sutherland speaks directly to the promising intersections among biology, physics and electronics that would come to drive artificial intelligence research (fields that are duly represented in the aggregated conference proceedings). From its inception as a discernible path of inquiry, the problem of artificial intelligence has always been considered an evolutionary one without a closed solution or a finite application (Minsky, 1959). The creative limit of AI research is a moving target, a perpetually shifting horizon of mechanization, but it has also been one with predictable social consequences, as the Dartmouth conference and related assemblies were invested in developing logic models that could map human cognition and produce thinking machines and effectively align human and machine processes. At the U.K. symposium, Lucien Mehl pursued a path as part of a subset of papers on the implications of AI research for industry; his research paper, "Automation in the Legal World" introduces mechanization as a tool to facilitate the retrieval of legal information, to search for relevant texts and jurisdictional precedents, but as it also points to the necessity to centralize information, it pushes further to consider how, pulling from these resources, a machine might provide a decision or be used for consultation within a highly specialized field of law (768). Though he provides the caution that we can have no "electronic judges," and "no machines to rule us" (778), Mehl's "law machine" requires the reduction and translation of legal principles into binary concepts and logical functions and necessitates a systematic analysis of existing and developing case law; as with all intelligence engines, it calls for an intermediate script that can negotiate between legal language, which is of course a human language, and machine language. Mehl's cataloging and retrieval system, a system he proposes might automate the more common set of legal arguments, is an early theorization of what would become governing industrial practice: the formal development of fully automated administrative systems with field-specific expertise. Historically, machines have been associated with the automation of various parts of the labor process (Moore and Woodcock, 2021), and the theorization and advancement of artificial intelligence have heightened this pursuit while raising the specter of technologically driven workforce displacement.

In 1961, MIT Professor Martin Greenberger organized the lecture series "Management and the Computer of the Future." Sponsored by the institution's School of Industrial Management (now the Sloan School of Management), the eight lectures were devoted to examining "the role and influence of computers in managerial decision-making" (MIT, 1961). In his contribution to the series, a paper on "Time-Sharing Computing Systems," John McCarthy (1962) speculated that computation could someday be organized as a (subscription-based) shareable public utility. McCarthy effectively outlined the foundations of distributed computing, and his speculative work was realized in the centralized hosting of business applications by IBM and other mainframe providers in the 1960s and through the emergence of network systems over the next decade (systems that followed from Bob Metcalfe's first Ether Network specifications, which he outlined in a 1973 Xerox Palo Alto Research Center memorandum). These collective efforts were ultimately expanded in the 1990s with the growth of the Internet, the rise of application service providers, the development of virtual private networks and the birth of cloud computing (a term used by Compaq in a 1996 analysis of the company's Internet-related business opportunities). McCarthy was not alone in his vision of a future grounded in and shaped by computation. Martin Greenberger had similarly anticipated the widespread application of computing in "The Computers of Tomorrow," a 1964 essay for *The Atlantic Monthly*. Reaching beyond the predictions of Vannevar Bush, who 19 years earlier penned "As We May Think" for the same publication, Greenberger prognosticated, "Computing services and establishments will begin to spread throughout every sector of American life, reaching into homes, offices, classrooms, laboratories, factories, and businesses of all kinds" (Greenberger, 1964).

The utility model of computing offered up by McCarthy and Greenberger gained traction throughout the 1960s, a decade dominated by mainframes, with a number of companies selling on-demand access to their resources: data storage, computing power and software solutions for accounting, financial planning and project management. The move into expert systems and away from general problem-solving (the primary focus of earlier AI research) was pioneered by Edward Feigenbaum during his tenure at Stanford. Feigenbaum took the helm of the university's Knowledge Systems Laboratory, where his research on knowledge engineering served as a lynchpin in launching commercial AI products. These systems and services constituted a new business model and were a source of competitive advantage among technology companies (including Bell Labs, Hewlett Packard, IBM, Texas Instruments and Xerox) and prominent research institutions (most notably MIT, Carnegie Mellon University and Stanford University). The preface to a 1982 United States Department of Commerce report calls out the rather limited achievements of early AI research—efforts that had been successful in solving "elementary or toy problems or very well structured problems such as games" (Gevarter, 1982) but fell short of the expert systems that would prove far more adequate in solving real-world problems. As one of its more speculative findings, the Department's overview of expert systems identifies growth opportunities in intelligent games, home entertainment and shopping.

Within these entities, expert systems, each built on a domain-specific knowledge base and designed to solve a particular set of problems, became the business case for investing in artificial intelligence. AI can make expert systems more situated (capable of perceiving and acting within their localized environments), flexible, autonomous and social (responsive to queries). Most sectors of society have developed parallel dependencies on expert systems, using AI to create value, to reveal the patterns in otherwise uncorrelated assets and to produce knowledge from information. Expert systems have become central to a number of industrial and municipal operations that include accounting, commerce, education, energy, engineering, finance, government, healthcare, law enforcement, product design and manufacturing, the military, real estate, transportation and urban planning and development. What began as speculative work in the mid-1950s that took place largely in university research divisions, research on computer reasoning, learning and natural language processing has matured into the expert systems that undergird corporate and institutional knowledge (Feigenbaum et al., 1988).

Research on knowledge-based systems was fueled in the 1980s through fierce international competition, as the United States, Europe and Japan sought to establish footholds in the information technology market. Japan's Ministry of International Trade and Industry launched the Fifth Generation Computer Systems project in 1982, a multimillion-dollar research partnership between the

Japanese government and the nation's computer industry that aimed to create a new generation of computers that would benefit from the gains made in artificial intelligence. The first gathering of the International Conference on Fifth Generation Computer Systems was held in Tokyo in October 1981 to outline the central themes of the work ahead, and reconvened in November 1988 to mark the country's progress. While the project did not yield significant breakthroughs in machine intelligence, it did manage to articulate the broader social impacts of AI as it set the international agenda (Pollack, 1992) and it spurred significant funding for computer science research in other countries throughout the 1980s: the collaborative European Strategic Programme for Research and Development in Information Technology (ESPRIT), Britain's Alvey program and the Microelectronics and Computer Technology Corporation (MCC), a research consortium of private technology companies with a campus in Austin, Texas. In response to the Japanese government-sanctioned effort, in 1983, the United States Department of Defense launched its own Strategic Computing program and spent more than $1 billion of federal funding over the next decade on research and development in computer design, artificial intelligence and a range of AI-based military applications that would leverage existing work on expert systems. The Department of Defense set the broad and ambitious goal of developing the machine intelligence technologies that would increase the nation's security and bolster its economic strength (Defense Advanced Research Projects Agency, 1983). In its initial planning document, the Department proposed the development of a broad spectrum of applications: autonomous vehicles, intelligent field assistants and sophisticated battle management systems. Many of these applications can be traced to Joseph C. R. Licklider's early theoretical work on machine intelligence, formalized in his 1960 essay "Man-Computer Symbiosis" and developed within the Information Processing Techniques Office (IPTO), a division of the Defense Advanced Research Projects Agency (DARPA) led by Licklider from 1962 to 1964. The IPTO was focused on harnessing computational intelligence to solve large-scale data handling problems, perform real-time situational analyses and monitor complex field operations—to produce automated systems for military command and control. The IPTO agenda was largely determined by a November 1961 report issued by the Department of Defense Analyses on the role of computation in the future growth of defense operations. Looking backward on intelligence research as a guide for the future, the report points to Claude Shannon's (1950b) seminal paper "Programming a Computer for Playing Chess" and proposes:

> It is evident that quite good programs could be written to aid a human chess player in making sure that his planned moves would not result in some serious loss through oversight. Military commands could undoubtedly benefit from similar assistance in plan evaluation.
>
> *(Belden et al., 1961, 11)*

Computers would be introduced into command operations not as human surrogates but rather to increase human capability—embedded in operations to analyze large volumes of information and generate variable "planning-to-command" cycles (Belden et al., 9).

Shannon introduced the chess machine as a discrete computational problem in search of a solution, and in doing so established a unique connection point between his earlier critical work on information theory and the theoretical and practical study of artificial intelligence. Shannon carried through with his proposal and built a chess machine that could handle various endgame scenarios. He also followed up his 1950 paper with a more popular discussion of his research in *Scientific American*. Shannon's detailed yet more accessible overview of the range of strategies that could be programmed into a computer opens with an acknowledgment that such a trivial problem might have a grander impact: "This problem, of course, is of no importance in itself, but it was undertaken with a serious purpose in mind. The investigation of the chess-playing problem is intended to develop techniques that can be used for more practical applications" (Shannon, 1950a, 48).

Time and again, play has set the field for artificial intelligence research and provided a ground for rigorous scientific inquiry. While play is commonly operationalized through structures and rules, it remains a cognitive activity and provides a space for agency; moreover, play folds into scientific research the persistent dependencies between science and society. Shannon's study of chess has rippled across other industrial fields, bringing with it his initial concern for understanding both the capacities of the human mind and the complementary capacities of machine thinking. When machine intelligence succeeds in the field of playable media, it does so with a balanced design imperative: to facilitate rather than dominate the experience of play. Research on edge cases, or scenarios in which AI systems do not perform as required or expected, has evolved from this early acknowledgment of the gainful attachments between humans and machines.

Playful Technologies

Roughly summarized, the first cycle of AI research followed the Second World War and was forged in military research centers and several major academic hubs (including MIT, Carnegie Mellon and Stanford); it was followed by a second cycle of research in the 1970s invested in the growth of expert systems and marked by a shift in focus from earlier work on general problem solvers to systems of domain-specific knowledge that could drive individual industries; and finally a third cycle that emerged in the 1990s focused on machine learning. What these eras have systematically brought together are the critical components of robust and accessible AI systems (the hardware, software, networks and interfaces) and the underlying programs that support natural language processing, knowledge base management, problem-solving and inference, all of which are pre-requisites for nurturing the symbiotic attachments between humans and machines (Licklider, 1960).

With a firm grasp on the historical foundations of the informatic layer of playable media, we can develop a preliminary outline of the links between play and intelligence. As part of this work, we can begin to map out the relationships between the two Tokyo-based intelligence initiatives that I have discussed earlier: the gathering on Fifth Generation Computer Systems and the release of the Sony AIBO. These projects represent two decades of intelligence research and are part of a longer arc that has seen AI move from experimental labs and into a range of playable products and usable development tools; in particular, deep learning, a subfield of intelligence research, has significantly advanced the commercial application of AI and has been given material form in an expanding array of objects and experiences.

The computer is not simply a cultural artifact (the product of a particular culture and embodying certain cultural precepts); it is also a site of cultural production. As the original inception of the computer is grounded in the needs of the military, solving the problem of fast and accurate information handling and decision-making, the present development of information technology remains bound to this legacy, although the computer has not remained an exclusively militarized machine as other industrial vectors have stepped in to support its development. While computer culture and military culture are intimately bound, with the former largely following the requirements of the latter (working with clear objectives, requiring speed and accuracy in information processing, and operating within highly structured systems of command and control), play can produce a number of viable industrial shifts to diversify how precisely computation is valued and developed (Whitby, 1986).

The game playing programs for chess and checkers that were pioneered by Alan Turing, Claude Shannon, Allen Newell, John Clifford Shaw, Herbert Simon, Arthur Samuel and a group of researchers at the Los Alamos Scientific Laboratory (work that spanned from the late 1940s through the late 1950s) served as models for testing machine intelligence and more fundamentally for writing programs that would appear to exhibit intelligence. These programs represented a shift in thought from hand-coded problem-solving to machine-learning algorithms and reflected a common understanding that the limits on machine intelligence were defined by the limits on human–computer communication (Newell et al., 1963). Moreover, these game environments provided useful spaces for decoding complex-problem-solving mechanisms, conceptualizing their nature and structure and looking outward to understand other intellectual and mechanical processes. The authors of the Los Alamos chess-playing program (written in 1956) were also members of the laboratory's nuclear bomb program, and their playful numerical methods were also central to weapon simulations: "To appreciate the need for numerical solutions, it is sufficient to recognize that the very nature of the Laboratory's objective—an atomic bomb—precluded extensive field testing" (Harlow and Metropolis, 1983, 132). This particular moment in the cross-industrial development of artificial intelligence is a striking reminder of the pervasiveness (and perhaps the perversion) of play in society, of play in the service of things that are decidedly not play (Huizinga, 1949, 2), of the transitory nature of play (as it moves toward an end state) and of the social manifestations of play that can be given concrete form.

The DARPA Gamebreaker Artificial Intelligence Exploration program was launched in May 2020 with nine research teams each assigned with two games: defined to test AI methods that could create game imbalances in video games and to consider the cross-industrial applications of these findings to Department of Defense war games, the project included a number of aviation, aerospace and defense technology companies (among them, Aurora Flight Sciences and Lockheed Martin) and several AI research labs. Aerospace and defense technology firm Northrop Grumman, which has significant investments in AI, was teamed with Hazard Software and Matrix Games and was assigned with *Command: Modern Operations* and *TORCS*. By applying AI modeling to an open-world video game, the Gamebreaker program seeks to "quantitatively assess game balance, identify underlying parameters that significantly contribute to balance, and explore new capabilities, tactics and rule modifications that are most destabilizing to the game" (DARPA, 2020). The overarching goal of the project is that those AI algorithms, when successful at assessing and then manipulating balance in commercial video games, might create similar imbalances in Department of Defense war game simulators—those modules used to train warfighters—and effectively translate game research into real-life military strategy (DARPA, 2020). Notably, defense AI operates within different parameters than many commercial AI systems and is designed to complement human decision-making during complex tactical operations and often in less orderly environments. Similar work on automated war gaming was undertaken by the RAND Corporation in the early 1980s, which folded the concept of meta-knowledge, or domain-specific knowledge of military strategic analysis into the development of the appropriate systems architecture. This contextual understanding of AI suggests that the characteristics of a system should reflect the unique properties of its domain (an understanding of institutional mission, conduct and operations) and its organizational behaviors—for example, the phased execution of a war plan should suggest how various actions go together and how certain decision points are reached or adapted (Davis, 1988). The context of applied AI is as important as its algorithmic architecture. In his review of RAND research on AI techniques in game-structured simulations of war, Paul Davis highlights the challenge of modeling heuristic decision-making when conditions are complex and uncertain, and he registers a certain skepticism that "generic knowledge engineers" and "generic software environments" would make for good localized problemsolving (Davis, 1988).

Expanding the Field of Intelligence

Technology is part of the human condition, and our bonds with machines are personal and social. Artificial intelligence has evolved through a series of porous intra-actions between humans and computers, and we continue to form new cognitive attachments with a material environment that is increasingly computational and more openly and organically communicative. These coevolutionary connections have changed society, influenced our politics and shaped our bodily practices (Hayles, 2012). If we consider the rational requirements of effective artificial intelligence, where reason is a simulacrum of both intelligence and intuition, we can recognize that technological agency is possible, and as something that resides outside of human agency (Giddings, 2005); we can refine our conceptualization of both the physical mechanics and the cultural logics of playable media by acknowledging that agency is produced through a series of complex relations between humans and machines.

While machine intelligence is transforming the global economy and revolutionizing employment, this is happening in the context of other pressing societal challenges. Anthony Elliott (2019) proposes that artificial intelligence is not so much an advancement of technology but rather "the metamorphosis of all technology" through the ubiquitous spread of its algorithms (xix). AI saturates our networked devices and experiences, and while it has impacted systems and organizations, it has also seeped into our personal lives. Situationally aware technologies have shaped our own situational awareness. Artificial intelligence provides the ability for technologies to adapt and self-learn and to draw digital corollaries from the human analogs of visual perception, speech recognition and language translation. I invoke the term playable media not as a simple extension of fun and games, but as a term that can encompass computationally driven media systems that use data to demonstrate reactive intelligence. While these systems may be fixed (but networkable) information architectures, embodied by a range of objects and experiences that are logical by design (a requisite of their scripting and programming languages), our encounters are more fluid, attached to larger narratives, incorporated into other actions and activities, and absorbed as part of the construction of self. Our transactional engagements with playable systems open up a series of transitional spaces (Winnicott, 1971) where we experience the dynamic interplay between the external realities of machine intelligence and its programmable materials and the internal realities of our own agency. Elliot (2019) suggests that "digital technologies form a key backdrop to how individuals learn to live between the lives they wish to have and the lives they encounter" (94).

In the chapters that follow, I explore a number of embedded AI practices that, to varying degrees, have closed the gap in human–computer communication and interaction and extended our understanding of cognition—practices that bring computation to bear on intelligent behavior. Playable media are grounded in the values of situated, embodied and enactive machine intelligence, and built through design frameworks that support such values and practices (Smart, 2018). Playable machine intelligence fosters a sense of connection and produces what we might consider functionally integrated (yet internally differentiated) systems that produce human agency through a form of mechanistic realization (Smart, 2018)—playing with and learning through technology. Human agents are written into the cognitive processing routines of playable technical systems, and those technical systems play an active role in shaping human agency. While early work on computational intelligence was focused on automation, a different vision has emerged of machine intelligence as a foundation for real-time interactive systems that can recognize and expand human cognition (Engelbart, 1962). Playable media are not designed to function as autonomous agents; rather they are designed to augment human behavior in different ways.

The focus of the following chapters is not on gaming or gamification but on the social and technical construction of playfulness and the evolving co-dependencies between play and complex artificial intelligence algorithms. Play can take place in games, systems and ecosystems, and increasingly machine intelligence has become a shared attribute of these distinct contexts, making them feel quite similar. More substantive than the expansion of game design concepts and elements into non-game contexts (Deterding et al., 2011), artificially intelligent systems represent a deeper evolutionary process: the application of problem-solving programming architectures, grounded in computation, studied and influenced by game design systems, to a broader field of praxis that is not always governed by explicit rules or oriented toward discrete outcomes (Juul, 2005). The spread of playful concepts and mechanisms in playable media has happened in tandem with the spread of physical computing—the proliferation of computational structures within the physical world—and the technical supports for ubiquitous computing environments. While playful technologies do not necessarily disappear into the background, as was the predicate for what Mark Weiser (1991) and his research colleagues at Xerox PARC termed "embodied virtuality" (Weiser, 1991, 98), their technical substrates are seamlessly designed as part of an "organic" system of environmental, geolocative, biometric and user-identification technologies (Nova, 2014). The underlying artificial intelligence frameworks and performative data networks of playable media form a bridge between the quantified self and the quantified world and provide the foundation for a digital data economy.

While computation has spread through almost every part of our social fabric (Golumbia, 2009) and has been used as a method of social organization and control, computational culture is not inherently oppressive. Algorithmic media are designed with a low tolerance for entropy, to reduce friction and produce satisfied systems; yet humans remain central to these playable and acutely responsive enterprises. Playable media are part of a larger landscape of artificial intelligence, and the following chapters lay out some of the unique permutations of intelligence. Far from a uniform praxis, AI is purposefully designed and embedded in distinct organizational forms, although these all move from the same functional premises of data collection and computational analysis and strive for perfecting the use of information. Machine intelligence has been enhanced by reinforcement learning and the ascendancy of autonomous algorithmic systems, and special-purpose engines have been supplanted by more general learning machines.

Playable media draw from a range of algorithmic tactics and, generally speaking, are designed to exhibit intelligence; yet from a purely functional perspective, some of these tactics are more adaptive and responsive to complex environmental cues and are realized as a form of evolutionary computation. The field of playable media continues to evolve, as a series of otherwise non-algorithmizable problems are inevitably solved. As my focus throughout this book is on systems, I have admittedly left room for other scholars to pursue more nuanced social impact research, including the influence of artificial intelligence on early childhood development in the context of the slow but steady growth of AI-first economies. Sherry Turkle (2021) suggests that "artificial intimacy" (a by-product of empathic machines) has lured us (both adults and children) into a false sense of attachment that, conversely, from an industrial perspective, simply treats us as things, as data targets.

Artificial intelligence is transforming the media and entertainment industries, revolutionizing content creation, marketing and distribution and accelerating demand side response; and big technology firms like Amazon, Apple, Facebook, Google and Microsoft have heavily invested in AI while diversifying their media hardware, software and content holdings. Algorithmic media analytics, decision-making and content generation are linked within these corporate ecosystems, and broad swaths of social, cultural, scientific and user data are channeled into playable media, synthesized and expressed through new physical and graphical forms and modes of interaction. Cultural theorist Lev Manovich (2018) suggests that the broad-scale adoption of computational methods has birthed a new set of "data-centered cultural techniques" (20) for encountering and acting within the world. Playable media need to be understood within this context as an industrial form capable of delivering a personalized user experience that also understands its user (as a series of inputs and metrics), but playable media also need to be understood more distinctly as performance-based problem-solving systems, where artificial intelligence is coupled to behavioral engineering and design to produce a number of intrinsic rewards. Playable media force us to engage with a question that historian Johan Huizinga (1949) actively chose to disregard: whether science can reduce play to a number of quantitative factors (Huizinga, 1949, 4). At the same time, the migratory patterns of both media and computation force us to reconsider the spaces of play—the spatial and temporal integrations of play with other practices. Even if the (operating) systems of playable media remain self-contained insofar as their software layers are hidden and cannot be re-engineered by casual players, they are commonly part of a networkable and connectionist AI paradigm, learning from larger data patterns that exceed their material structures. The expanded material infrastructures that support playable media, and the labor associated with acquiring, storing and processing data remain hidden to sustain the illusion of automated intelligence (Apprich, 2018). Play can be conceptualized as a medium insofar as it is attached to lived experience and can be organized as a structured encounter within a unique discursive context (Rodriguez, 2006); and media can be played when they extend such an invitation to act through a series of structured resources and conditions. Play is functionalized in playable media, not necessarily as a mechanism of socialization and control, but because of the very rational requirements of programming, computation and information architecture.

Sony first announced the development of OPEN-R in 1998 as a new architecture for home entertainment robots, demonstrated on a four-legged prototype (the foundation for AIBO development), that would allow interchangeable hardware and software modules and support multiple physical configurations and expanded applications (Sony, 1998). Four years later, Sony released the OPEN-R Software Development Kit (SDK) and opened up its architecture to developers. With these specifications in hand, the AIBO was essentially unlocked for building customized software programs. In 2019, Sony introduced the developer program, accessible through a proprietary Web API for the ERS-1000, the fourth generation of its companion robot, along with new programmable resources such as the ability to potty train AIBO with the "toilet is here" command (although the robotic puppy does not urinate or defecate and only eats virtual treats). As an advancement on the earlier SDK, the more recent developer program is also accessible with Sony's Visual Programming tool, which allows for drag and drop modifications to AIBO's behaviors and movements. By repeatedly opening up its platform, Sony has expressly aimed to accelerate noncommercial robotics and artificial intelligence research. But what does it mean to open up a playable media system?

That is the question that drives this book, and my goal is to illuminate playable media as institutional frameworks, as structured objects, as process-oriented experiences that remain connected to broader industrial and informatic exchanges, and as responsive computational systems with intentionally designed forms and functions that can involve different types of actors. Playable media may share a common set of technical and performative properties, but they can serve distinct user groups and have distinct purposes. As procedural systems, as unfolding works and as personalized journeys, playable media contribute to our sense of self, yet they are dynamized by their internal AI scripting. This book examines a series of fixed assets and ecosystems, but it is meant to encourage broader intellectual and creative considerations of the cultural role of artificial intelligence; it is intended to open up the field by opening up a series of playable objects and systems. Playable media is an expansive field that has emerged from a more generalized playful mindset and an evolving series of social contracts that define where play happens and what activities might be made over by the conditions of play. Playable media emerge and are codified from these conditions, techniques and technologies (Stenros, 2014). My attention to the computational design processes that undergird playable media is an attempt to call out the technical aspirations and the discursive formulations that continue to shape user-centered machine intelligence. Artificial intelligence research has yielded a bounty of cognitively grounded functional applications, and AI research has become a critical part of software studies, as it has changed the fundamental way that code works and how software performs in the world (Kitchin and Dodge, 2011). AI has reshaped a number of contextual software practices—in commerce, education, entertainment, finance, governance, healthcare, manufacturing and transportation—and the media artifacts associated with those practices. Artificial intelligence has expanded the field of playable media by galvanizing the production of new code objects, new embedded relationships and new forms of work, leisure and knowledge that consistently engage with playable resources.

Playable media are communication systems. They rely on internal data exchanges, and they combine and integrate a number of distinct media forms to trigger a range of semiotic activities—the external human-centered expressions of their internal systems of exchange. Playable media are realized through their underlying "technical media of distribution" (Elleström, 2020, 3)—the material and immaterial processes that allow them to harness networked data and establish human connection. Playable media translate data into action by systematically linking their algorithms to material outcomes, and in the chapters that follow, I trace this work through a number of field-specific case studies.

Playable media can be helpful or harmful, but their disposition is largely linked to the conditions of their ownership, production and operation, as well as a number of functional questions about the exchange value of our data and the personal return on our investment. We might ask how much play each system has—how much slack there is and how much opportunity for action. As I have suggested at the beginning of this chapter, the term playable media is meant to invoke a sense of freedom, but that description may be too conceptual. A more functional approach will allow us to consider how, precisely, these adaptive or variable systems work and how they learn from our engagement.

2

EXPRESSIVE INTELLIGENCE

Modernizing the Video Game Industry

The algorithmic foundations of the video game industry lie in the computer sciences, while video games have in turn provided important benchmark problems for testing and expanding computational intelligence. Game AI is a subset of broader intelligence research that has been shaped by the steady evolution of computer software and hardware and conditioned by the entrepreneurial goals of the video game industry. AI has made its imprint on video game development, as a process and as a design solution, and has been a powerful force in moving the game industry and other intelligence industries forward. Game AI is part of a larger software framework (a game engine) that binds the state of production (software as a development tool) to the state of play (software in its execution). Game AI is expressive in that it ties graphical functionality (the matter of game engine architecture and such processes as decision-making and pathfinding) to textual design (the matter of game art and assets), playability and engagement; and game AI is expressive in that it conveys a range of ethical precepts and cultural values that are deeply embedded in the built information environment (Star, 1999) and part of an industrial impulse to create more calibrated gameplay experiences by addressing player sentiment (Ambinder, 2011). These interlinked graphical, textual and procedural logics (Wardrip-Fruin, 2005), bound through real-time rendering, form the primary structures of play. Game AI is more than a mechanic; its actions are more expansive, assuming multiple forms and functions within a game world. Game AI is also expressive in that it is shaped by its computational context—the historical, cultural and economic conditions that influence the hardware and software shared by computing and gaming (Wardrip-Fruin, 2009). This chapter provides an overview of the development of video game AI and identifies the primary system models, their applications and their impact on play. Game AI is simply one implementation of AI, part of a broader field

DOI: 10.4324/9781003225072-2

of machine intelligence that has been developed to complement many distinct industrial vectors and many unique operations. Within the video game industry, game AI is part of the unique design considerations of a particular development workflow, and the end goal of producing a believable intelligence system needs to be balanced by the more immediate goal of producing a system that is easy to use by a range of design personnel (including level designers), and where parameter creep can make a system too complex for every system user (Isla, 2005).

The computing research community has interacted with the video game industry in ways that have, not surprisingly, foregrounded computation: to solve the technical problems associated with existing game genres (e.g., pathfinding in complex open-world games, the networking protocols for multiplayer games), to explore the application of existing AI research such as machine learning to games, and to expand the application of game-based intelligence systems to other industrial domains such as healthcare and aerospace, where AI can be leveraged for training, predictive analytics, supply chain management and generative design. But this approach to computing research privileges the science of computation over the relational nature of any given medium and often fails to embrace the computational flexibility and expressivity of playable media. Computer science is not a natural gateway to media-centric AI research or meaningful computation. Yet beyond teaching machines to learn, reason and act independently, AI has, on occasion, shaped game aesthetics and expanded the form of play; and there is value in considering how a number of distinct research disciplines can interact and necessarily contribute to the evolving practices of the video game industry and expand our understanding of what is possible in games.

While the foundations of the video game industry lie in the computer sciences, as the game industry has evolved, it has pushed back and challenged many of those core principles and techniques, raised new questions, and drawn from research in other domains (Lucas et al., 2013). The contemporary development pipeline reflects this admixture in its artists, animators, composers, level designers, writers, testers, engineers and programmers, and in the ongoing refinement of these roles. AI programming has become its own subfield and specialization in the video game industry, with studios looking to recruit programmers skilled in what are already deeply established game AI system design patterns; while AI controls many aspects of gameplay, and is revealed most forcefully in the actions of non-player characters (NPCs), it is also part of the design and development process. On the production side, the push toward intelligence is more pointedly about computational methods for optimization, learning and control, or more broadly for improving game performance and realizing greater efficiencies in game development. AI is a vehicle for a series of interlinked interactive processes: problem-solving, decision-making and worldbuilding.

While video games have been a fundamental testing ground for artificial intelligence, we should also examine whether video games have in turn shaped the development and implementation of AI; larger game developers have expressed an interest in AI research, and their industrial paradigms are undoubtedly influencing the field, where AI is viewed through the lens of game (application) space. AI is changing the field of game development, and in turn, game development is changing the field of AI. We see this in the sharing of industrial resources—the cross-industrial application of game systems and AI subsystems to build virtual environments and run scenarios within them, and the cross-industrial deployment of game labor, porting the technical expertise of video game development to non-game or game adjacent applications. The metaverse is being built with many of the same systems that were developed within and are driving the video game industry, and these systems are being remapped to new use cases by many of the same technology laborers who have engineered and iterated them. Epic Games, Roblox, Tencent and Unity Technologies are illustrative of the commitments that game developers (and those companies that own game studios as part of larger investment portfolios, and those that publish and license software) have in building more expansive immersive platforms that can accommodate new markets and services and new methods of monetization. Epic's *Fortnite*, a 2017 game built on the company's Unreal Engine, is not only simply a play space but also a creative space with functional design tools, a community space and an event space; similarly, Roblox has focused on building its immersive world with existing game technologies and through nascent forms of "metaversive" sociality that include making games, playing games and building community. By pursuing a strategy that integrates AI into every commercial vector, where AI may have a function if not a form, these companies are able to collect massive quantities of information about their clients, whether through social AI (intelligent agents) content AI (recommendation engines), platform AI (AI as a service) or video game and medium-specific asset-based AI (AI-inscribed content).

Game AI as History, Theory and Practice

AI has various uses within the video game industry: as a design tool to automate asset creation, level design, and worldbuilding and speed up the overall development process, as a quality assurance or playtest bot (used to identify bugs and coding errors or perform stress tests), and as an analytic tool to understand player motivations. These uses expand on what is the more common critical attention to AI in video games: as an agent associated with non-player characters (often designed as human-like opponents or companions that can execute rational oppositional or cooperative behaviors) and as a more general category of believable agent or agentive other that can provide support, challenge or conflict. Video games provide a number of unique challenges for AI, including developing non-player characters, creating procedurally generative systems that can construct game content in real time, and defining and managing worlds, resources and agents that can learn from and adapt to players by analyzing their in-game behaviors. Many of those cognitive insights and solutions have been adapted to solve real-world problems, heightening the relevance of AI games research—a field that has coalesced over the last two decades following the first conference on Artificial Intelligence and Interactive Digital Entertainment (AIIDE) and the first IEEE (Institute of Electrical and Electronics Engineers) Symposium on Computational Intelligence and Games (CIG) in 2005. These gatherings of academic researchers and industry developers predicted the importance of AI in digital entertainment both behind the scenes, as a form of data management, and in front of the scenes, explicitly embodied in machine-generated character classes, assets and environments; and they were a natural outgrowth of more than 30 years of development work in the video game industry that advanced computer hardware and software. The 2005 AIIDE conference featured presentations from Bing Gordon (Electronic Arts), Damian Isla (Bungie Studios), Craig Reynolds (Sony Computer Entertainment) and Will Wright (Maxis), and pointed to significant changes in gaming—support for player choice, high-fidelity open worlds and less linear sequencing—that were impacting the development process. These changes were linked to advancing AI in particular directions, to realize greater flexibility and expressivity and generate multiple forms of procedural response: procedurally generated content and contextually generated character awareness in independent agents. These technical solutions were also linked to the matter of visual representation and spatial coherence and competence, tethering the work of programmers to the work of animators. Gordon (2005) suggests that developers should openly articulate what their AI is doing: "Game AI must be acted out and seen." Players must be able to understand the play concept, visualize the flow of information between environmental situations and behavioral responses, and be given clear evidence of the impact of their actions in the game world; these signs of intelligence create meaningful play. There are, of course, many applications of game AI that do not involve opponents or non-player characters; the principal features of built systems, of data management, simulation and automation, can be understood as less embodied forms of computational intelligence, although they are conveyed through material form (and quite obvious when they fail to work). Much like the material systems in which it is embedded, artifacts that include video games, artificial intelligence speaks volumes (not literally, of course) about its historical, social, cultural and economic context (Sotamaa, 2014); indeed, if we read AI only from the perspective of its game-based applications, we can see the shifting nature of the video game industry, from material systems to information systems, the constant interdependencies of hardware and software, and the social shaping of labor and production.

Research on artificial intelligence in video games is indebted to the contributions made by Alan Turing, Claude Shannon, Arthur Samuel and others, all of whom used familiar games such as chess and checkers to demonstrate novel programming ideas (evaluation functions, minimax procedures and reinforcement learning); in doing so, they defined many of the algorithmic intelligence processes that would drive future AI research (Fogel, 2005). Yet the path forward, as outlined by much of the scholarship presented at the 2005 IEEE CIG meeting, productively expanded machine learning beyond the programmable limits of traditional AI to suggest a new horizon of self-learning and adaptation, a turn toward computational intelligence and evolutionary computing methods that might "hybridize" with human knowledge (Fogel, 2005).

Furthering the work of these two professional associations, the May 2012 Dagstuhl Seminar on Artificial and Computational Intelligence in Games gathered more than 40 researchers at Schloss Dagstuhl (Leibniz Center for Informatics) in Saarland, Germany, to discuss a number of interrelated strands of applied AI games research: computational narratives, pathfinding, player modeling and procedural content generation. As the organizers and editors (Simon Lucas, Michael Mateas, Mike Preuss, Pieter Spronck and Julian Togelius) of the conference and its published proceedings note, "Almost every AI and CI [computational intelligence] technique can be seen as search in some way: search for solutions, paths, strategies, models, proofs, actions, etc." (Lucas et al., 2013, viii). Video games provide a foundation for identifying complex problems and acting through them; they provide a fertile ground for algorithmic thinking and testing. The majority of the problems being worked through at the Dagstuhl Seminar were localized to game environments, which is not surprising, given the gathering was focused on the study of games. The studies, when read collectively, suggest the importance of building a bridge between academia and the video game industry and highlight the possibility of producing better gameplay experiences with the support of advanced AI, but they do not push further into the broader industrial application of intelligent agents. Yet a study on player modeling led by Georgios Yannakakis broaches the grander topic of human–computer interaction to consider how developers might exploit the computational means for modeling player behavior, cognition and affect to create more believable agents (Yannakakis et al., 2013); we can see this as an early effort to consider how intelligent agents might perform in other fields.

The history and development of artificial intelligence have particular points of industrial confluence; notably, the exponential growth of computer hardware and networked communication since the 1990s has changed the field of both computing and gaming, making it possible to simulate complex physical environments that demand greater computational intelligence. Standard labor-intensive scripting and authoring methods have yielded to a variety of computational intelligence techniques, research reflected in both the CIG and AIIDE symposia (Miikkulainen et al., 2006). Video game development is also tied to advances in artificial intelligence and the demands associated with negotiating multiple system agents, integrating discrete sensory inputs, reacting to player behaviors and generating dynamic responses in (and modifications to) a simulated physical environment. AI techniques excel at these tasks and can manage both the specificity and the plasticity of a numerically constituted and algorithmically governed domain (Miikkulainen et al., 2006). Sophisticated game agents can display visibly intelligent behavior as they learn, adapt and act, in real time, to new situations.

One of the earliest and fullest conceptualizations of AI in digital games, a 2001 *AI Magazine* article by John Laird and Michael van Lent, points to the promise of interactive computer games to realize robust, human-level computational intelligence. As a natural extension of the game industry's push toward realism and within a developmental framework of applied AI, graphical realism would be matched to behavioral realism, and simulation would be enhanced by the sensory logics of physical and social interaction (Laird and van Lent, 2001). While the behaviors of NPCs in early video games were scripted or relied on simple rules, the programming vocabulary of game development expanded as AI research expanded and notably as hardware power also expanded (Yannakakis and Togelius, 2018). The state of AI practice shifted significantly in the mid-1990s; 3D hardware acceleration opened up the opportunity to develop complex physical environments which in turn placed greater demands on data management (in emergent game genres such as the first-person shooter).

Game AI has been continuously shaped by a range of institutions, agents and contexts; and the most successful couplings of play and intelligence seem to be built with an open acknowledgment that current technical limits will necessarily shape the design process. The histories of video games and of game AI are interwoven and interdependent, yet as Georgios Yannakakis and Julian Togelius (2018) note, these histories also carry with them their own legacies that make them often difficult to synthesize; modern video games have evolved from earlier game designs, and some of the earliest basic design patterns (platformers, first-person shooters and real-time strategy games) that were established in the 1980s and 1990s with only nascent AI may not provide the best models for understanding the potential of more fully developed AI or the furthest horizons of computational creativity (Yannakakis and Togelius, 2018).

Industrial Benchmarks of Game AI

Before its widespread application in the home console market and in next-generation video game systems and AAA titles, rudimentary game AI was a feature of arcade machines, designed to create a sustainable challenge and implemented as an oppositional force through scripted rules, actions and a degree of random decision-making. Arcade-bound AI was designed to keep players engaged, while attached to a series of limits on play such as the number of lives and a widely adopted level-based paradigm that introduced progressive difficulty, both of which kept quarters flowing into machines (Tozour, 2002a). AI was hard-coded as a series of stored patterns in 1970s' arcade games, and in many cases influenced game design. The influence is obvious in *Pac-Man* (released in 1980), a classic arcade game with hard-coded patterns that are distributed across individual enemies; the game's various ghosts seem to act independently but are tethered to a single AI system. The handheld market of the 1970s, which included Mattel's *Auto Race* (1976) and *Football* (1977), Parker Brothers' *Merlin* (1978) and

Milton Bradley's *Simon* (1978), largely featured programming directly designed into its hardware. *Auto Race* was built from an LED display and a compact circuit board; the game's programming was written in an assembly language and stored on a modified calculator chip. Most of the decade's handheld games were written in assembly language (a low-level programming language); and while they featured only rudimentary code rather than advanced AI, the majority of these microcomputer-controlled games were situated as intelligent opponents (although one of the key features of *Simon* is the inclusion of social multiplayer modes). The ad copy for *Merlin* challenges players to outsmart the machine, and the game manual describes Merlin as "very talkative" and cautions, his "computer brain is made of many delicate electronic parts" (Parker Brothers, 1978). These handheld games had integrated programming and hardware; they were objects with fixed components that could not be readily iterated or advanced, and their intelligence was limited to a number of stored patterns and code sequences, yet those limitations were shaped into a series of narratives that consistently foregrounded the brain power of human-like machines.

The emergence of the home console market and the processing wars that followed shifted game-based AI research toward intelligence and fidelity, with genres that emphasized real-time strategy (RTS) and called for real-time visualization. Historically, game AI research has lagged behind the industry's commitment to building 3D engines to render advanced graphics, and AI work has been consigned to later stages of the development process, but following the growth of the RTS market in the 1990s, those responsible for game AI have been involved earlier in the development cycle. As a result, AI systems have been more firmly integrated into game design and game engine architecture.

Modern game-based AI has developed in tandem with the video game industry's turn toward game engines, which manage the computational tasks that are central to advanced digital media production pipelines. A number of hardware advances, which include the shift to multicore processing in PC and gaming platforms which began in 2005, forced significant changes in engine design and led the industry to pivot toward parallel development to increase computational capacity. This pivot allowed developers to take advantage of multicore processors and chips, and allocate more computational memory to AI tasks while allowing these tasks to run in parallel with other game functions. The functional parsing of runtime elements and the move from serial programming to data parallelism led to a common multithread engine design with modular component-driven subsystems (Freedman, 2020).

The most prolific game studios have navigated these shifts by producing their own core engines, while others have turned to several of the major licensable 3D authoring tools, including Unity and Unreal. Traditional engine-based histories of video gaming situate *DOOM*, developed by id Software and released in 1993, as an origin point; and we can track the steady integration of AI subsystems into game engine design from that point forward. The birth of game engines can also be linked to the development of independent middleware modules in the late 1980s and the early 1990s that were designed to handle advanced video game graphics. As game applications and runtime environments became more complex, developers built increasingly robust engines to manage a wider array of tasks, and artificial intelligence emerged as a distinct subsystem that could be integrated with graphics, physics, audio and other game components (Freedman, 2020).

The two dominant commercial game engines, Unity and Unreal, are equipped with AI tools for modeling the behaviors (awareness and decision-making) of non-player characters and other in-game entities. The Unreal Engine's intelligence assets include behavior trees, AI perception and a system for querying the environment (EQS). Unity machine-learning agents (ML-Agents) are open-source assets available through Github that combine deep reinforcement learning with imitation learning to create intelligent behaviors for non-player characters and to build complex AI environments to benchmark new algorithms and methods. While many game developers deploy turnkey solutions such as Unity ML-Agents for their AI subsystems, those studios that create their own engines commonly build their own AI integrations. In both pathways (deploying prebuilt solutions or designing custom ones), machine-learning agents can help a studio realize more efficient game-level design, reduce overall production costs and streamline communication among members of the development team. Advances in 3D hardware acceleration (which are tied to building faster CPUs) have increased the amount of processing power apportioned to the work of AI and have led most game developers toward deepened dependencies on AI, while the codework that comes with building emergent behaviors has been reduced by prebuilt intelligence systems. Game labor has been radically transformed by proprietary AI systems and video game engines, as have the methods of team-based programming and design, the mechanics of distribution and the associated field of localization (adapting a video game property, its assets, spoken language and text, for multiple regions).

AI systems can learn to independently generate content and, in advanced applications, automatically build entire game levels; and AI systems can be coupled with other software frameworks to radically alter the game development pipeline. In February 2021, Epic Games announced its MetaHuman Creator, a software platform that uses the studio's proprietary Unreal Engine to create realistic looking and moving virtual humans with detailed facial animations. Epic's investments in high-fidelity digital humans were advanced by the company's 2019 acquisition of Serbia-based 3Lateral, a move that pulled the technology for crafting virtual humans closer to the Unreal Engine by adding cutting-edge research on volumetric facial capture and facial rigging to Epic's portfolio. While the MetaHuman Creator does not rely on artificial intelligence *per se*, and is instead focused on the physical aspects of character creation, these assets are primed for intelligence uses, as non-player characters driven by AI subsystems, or as intelligent agents in a wider array of video game-adjacent interactive environments. Yet Epic's current End User License Agreement limits using MetaHuman technology "for the purpose of building or enhancing any database or training or testing any artificial intelligence, machine learning, deep learning, neural network or similar technology" (Epic Games, 2022). Since its launch, MetaHuman Creator has been successfully coupled with an AI voice actor platform developed by Replica Studios, enabling artists to create AI characters that look and sound like humans. Replica's AI model learns by copying the speech patterns and intonations of real voice actors; using these digital assets, the studio has built a library of more than 40 AI voice actors that can be exported into Unreal, Unity, NVIDIA Omniverse and other 3D development platforms, and synched with the context-specific facial animations of in-game characters. In November 2021, NVIDIA announced Omniverse Avatar, a product that connects the company's research in voice AI, computer vision, natural language processing and simulation (NVIDIA, 2021); interactive characters created in the platform can see, speak, converse on a wide range of subjects and understand naturally spoken intent (NVIDIA, 2021). Pursuing a similar strategy, Roblox purchased Loom.ai in 2020, acquiring the company's real-time facial animation technologies (that combine deep learning, computer vision and robust visual effects) to advance the platform's own realistic 3D avatars. South Korean software developer Plask launched its self-titled AI-driven animation tool in January 2022; the browser-based tool, which features integrations with Maya, Unity and Unreal, uses machine-learning technology to extract full-body motion data from video footage and retargets it to 3D characters. While these parallel developments in character building may seem outside the immediate purview of game AI, as they push their avatars into other platforms and experiences, they emerge quite literally from game industry investments in real-time 3D worldbuilding; and of course, these technologies reveal several of the ways that AI has been used to develop in-game content, simplifying game-oriented motion capture and character animation and other 3D visualization processes.

These investments continue to socialize the technologies of game development among players and developers; as users demonstrate new products, they also build a shareable knowledge base and deepen their attachments to proprietary platforms and services. Roblox (released in 2006) was launched as an online game platform, as a space to bring people together to play games made by other developers (using the proprietary Roblox Studio engine), while Epic and Unity have developed robust video game software ecosystems with parallel sharing economies and developer hubs. NVIDIA, Epic and Unity are regular partners, with NVIDIA hardware powering the engine-based pursuits of both software developers and in the case of Epic, providing custom Unreal Engine branches for NVIDIA technologies as open-source resources on GitHub. This sharing economy, while focused primarily on development tools, has added material depth and relevance to proprietary hardware. NVIDIA's deep-learning super sampling (DLSS) technology, which relies on the company's RTX line of graphics cards, has allowed Unreal developers to leverage accelerated graphics rendering using the chipmaker's proprietary AI algorithm. DLSS technology can boost video game frame rates by rendering at a lower resolution and using deep learning to provide real-time image upscaling; as with other integrated hardware and software solutions, DLSS has a recognized material consequence within the game development pipeline.

While not an identical technique, DLSS has much in common in its transcriptive relationship to visual culture with the imaging principles of computational photography and digital processes such as focus stacking, high dynamic range (HDR) and pixel binning, which can be used to enhance images and reduce noise. Apple's Neural Engine (ANE), introduced to the iPhone in 2017 as part of its A11 processor, is a type of neural processing unit that accelerates the device's neural network operations. The ANE effectively expands and mobilizes a range of visually oriented machine-learning applications and allows the iPhone camera to leverage several forms of advanced image processing and analysis to support, among other tasks, facial identification and object recognition. The ANE has significantly enhanced the iPhone's photo suite and the volume of data that can be gleaned by its camera. Though these developments in camera-based technologies are beyond the localized field of game AI, they illustrate the convergent pull of machine-learning algorithms within material hardware systems (phones and cameras) and across visual culture. These technologies also remind us that AI operates at every level of representation; it defines worlds, characters, assets and images, and uses data to manage both visual and performance-based behaviors.

Performance-Based Intelligence

Beyond development, localization and automation, modern video game intelligence is designed for in-game performance, notably in combat, pathfinding (getting from point A to point B) and procedural content generation. Advanced 3D games use AI systems as the foundation for intelligent movement, collision detection, and for establishing finite states (a fixed set of states or actions that define what a character can do, and that can only be performed sequentially) that regulate the behaviors of non-player characters. This foundation has set limiting conditions for more expressive games that try to advance AI in new directions—to explore new forms of communication, develop new models of interaction or promote empathetic behavior. At a May 2021 corporate strategy meeting, Sony CEO Kenichiro Yoshida announced a partnership between Sony AI, the company's artificial intelligence division, and PlayStation to develop more responsive computer-controlled characters: "By leveraging reinforcement learning, we are developing Game AI Agents that can be a player's in-game opponent or collaboration partner" (Yoshida, 2021, 38). In February 2021, Warner Brothers Entertainment secured a United States patent for the Nemesis System, an AI mechanic developed by Monolith Productions for its *Middle-earth* games (*Shadow of Mordor* and the sequel *Shadow of War*, released in 2014 and 2017, respectively) that allows non-player characters to evolve in reaction to game events. The patent, filed as "Nemesis characters, nemesis forts, social vendettas and followers in computer games," describes a system of procedurally generated non-player characters whose parameters and hierarchies can change in response to their interactions with players (De Plater et al., 2016). In 2020, Microsoft launched Project Paidia, a research collaboration between the company's game intelligence group at Microsoft Research Cambridge and game developer Ninja Theory (part of Xbox Game Studios). The joint research project is focused on using reinforcement learning to develop non-player video game agents that can learn to collaborate with human players. This repeated emphasis on refining non-player characters and validating their agency suggests that the modern video game industry is focused on one tenet of player engagement and one common point of failure. Non-player characters serve both a narrative and a tactical function within a gameplay environment; they are gestalts of game development, design and play, embodiments of hardware, software and programming, and validations of system performance.

AI has prescriptive power in an industry driven by efficiency yet prone to performance issues, and by the nature of the distributed work of AI in both game development (as a tool) and play (as a process). That power is illustrated in the work of modern intelligence systems, and, in particular, the dominant systems developed or adopted by large and mid-size game development studios, including CD Projekt Red, Creative Assembly, Take-Two Interactive and Valve. While these studios are just a small sampling of a much larger industry, their work highlights the steady post-arcade market growth of AI over an arc that has included both significant breakthroughs and costly failures (notably, CD Projekt's *Cyberpunk 2077*, which had a disastrous launch in 2020). It is not my intent to provide an exhaustive inventory of modern video game AI. Rather, the central goal of these case studies is to understand AI algorithms in relation to their material outcomes—to tease out their formal influence and, by extension, their impact on gameplay by examining a number of proprietary systems and game titles. By pulling together several noteworthy titles, games that have been lauded or lambasted in the popular press specifically because of their AI systems, we can begin to conceptualize the performative value of game intelligence. Machine-learning algorithms are intended to be invisible gameplay supports, but their success or failure draws our attention to the material elements of each game—from satisfyingly challenging opponents to disastrously broken environmental mechanics and physics. The playable values of playable media are tied to functional, balanced and contextually motivated AI. There is an admitted bias here on the AAA titles produced by major game studios, as I am interested in the relationship between machine intelligence and worldbuilding and the physical imprint of AI. That interest will become clearer in the last chapter of this book, which considers worldbuilding across a number of distinct industries—industries that are bound together by a number of common tools. AI is, of course, a feature of both mobile and console games, and I have privileged use cases where it is bound to a game world and character-based environmental mechanics. In popular turn-based or puzzle-solving games, AI functions as a fairly disembodied mediator, a randomizer, and an opponent, and there is no precept that it is something other than machine programming. AI functions less as a fantastical projection in turn-based, puzzle-solving or simple shooter games, and more as a necessary element of solitary play.

The Aesthetic Vocabularies of Game AI

In the same way that video game engines are shaping game and non-game industries alike, as film and television production is transformed by the engine-based logics of virtual production and the new shooting methods, new hardware systems (such as LED walls and volumes) and new workflows that support real-time 3D visualization, artificial intelligence has established a new foundation for contemporary game development that has impacted how games are designed and developed, and is influencing the visual vocabularies and methods of other media. Virtual production pulls together the methodologies of several fields, including filmmaking, 3D graphics, computational photography and engine-based rendering. Game engines are influencing every aspect of the film and television production pipeline, including previsualization, and advanced in-camera visual effects are complementing what was once the purview of post-production. Directors and cinematographers work in tandem with digital imaging technicians and leverage game technologies throughout their respective production processes to seamlessly integrate physical and virtual assets. In this turn toward engine-based production, AI can be used to support real-time body tracking, facial mapping and 3D animation, and those supports are changing the visual vocabularies of film and television. AI is also aiding the internal data analysis of subscription-based streaming services, and more acutely aligning supply with demand. Platforms that include Cinelytic use AI to provide predictive analytics to film industry clients; Warner Bros. Pictures began using the tool in early 2020 to guide the studio's marketing and distribution decisions. AI has various uses, as a visualization tool, as a production support, as an analytics and recommendation engine, and embodied by formally constructed and animated character assets. Each of these uses is shaping the aesthetic vocabularies of film and television, and playing a determining role in what can and should be produced. Within the virtual production pipeline, machine learning has its most obvious applications in visual effects, in the production and animation of virtual assets and in performance capture. Digital Domain's proprietary Masquerade system uses machine-learning algorithms to produce high-resolution facial motion data from limited low-resolution source data, a process used to animate the computer-generated Thanos in the 2018 film *Avengers: Infinity War*. With the release of Masquerade 2.0 in 2020, Digital

Domain announced that it had optimized its facial capture system to expand into other digital character applications, including next-generation games, episodic television and commercials (Digital Domain, 2020). As this detour through virtual filmmaking suggests that, while AI is bound to the principles of computational modeling and functions as a technical strategy, it also shapes the aesthetic vocabularies of visual media; AI may be a data management tool, but it is also fundamentally expressive.

Engine-based AI links intelligence to action in its both virtual and physical applications; as an interoperable software architecture and a common foundation, machine intelligence undergirds systems, experiences and objects that share many of the same behaviors and potentials. AI allows platforms, bodies and technologies to work together (Ball, 2020) and enables us to build persistent models of the information world, but AI is developed in tandem with hardware and software, and in the context of more sweeping social, political and industrial realignments. Modern game-based AI systems have been designed with two central procedural purposes: to solve the algorithmic complexity associated with real-time visualization and to achieve the illusion of behavioral intelligence. Game-based AI is used to build a sense of cohesive realism (in sight and action), to create legibility and intelligibility, rather than to produce game-breaking outcomes that might interfere with play. Players are invoked as actors, but they are not regarded as individuals or protagonists who might alter a game's processes. Games have rules and are designed as hermetic systems (with the exception of those intellectual properties that are distributed through open-source models, or offer end-user modification or development kits). To realize the expressive potential of AI, we need to look beyond the bounded spaces of games—spaces that are literally authored and scripted—to consider how machine learning has redressed the individual in more expressly open and unpredictable cultural spaces as a measurable and knowable reactive agent. I start down this path in the next chapter. Real-time visualization is enabled by engine-based game design, development and execution, and these same engine-based supports are driving changes in other non-game media industries, where they are reproducing many of the same formal outcomes.

AAA video game development has advanced through a series of technical achievements and failures; the steady improvements in graphics hardware that support photorealistic gaming visuals have been matched by parallel advancements in game engine systems and artificial intelligence subsystems that support increasingly uncanny character performance. Together, hardware and software have advanced the art of simulation, and the standards and practices of video game simulation are steadily influencing the aesthetics of simulation in other industrial sectors. Across industries, simulation is being driven by the 3D visualization architectures that have coalesced in video game production and the studio-bound technological imaginaries that have sustained these efforts. The pursuit of hyper-realism in AAA game production is influencing visuality in non-game media and is being furthered by certain assumptions about the inevitable alignment and seamless integration of the physical and the virtual world. Photorealism and behavioral realism have developed in tandem as a unified computational problem to produce more natural and reactive environments with greater emotional fidelity; the challenges of creating a highly dynamic and involving world have been met by new graphics standards and new physical algorithms. Yet AI can fail; it can be too overpowered or too unintelligent. Poor AI systems can produce imperfect gameplay—experiences that are either too predictable or too unstable.

The Functional Vocabularies of Game AI

AI has been a standard feature of the video game market since its infancy, popularized as part of gameplay structure with the rise of opponent-based arcade games in the late 1970s. The classic fixed-shooter game *Space Invaders*, released in 1978, is designed with a fairly rudimentary intelligence system; stored patterns are used to simulate random enemy movements that are pre-programmed into the game. AI has become more central to the gameplay experience with the rise of multiplayer networked games that require the monitoring of an increasing number of potential game states, and in simulation games such as *The Sims*, released in 2000, that run on need-based formulas. In part, AI has moved forward (happens earlier) in the game development pipeline, and has become a critical part of game design. AI requires that game developers give more thoughtful attention to how they allocate processing resources and the amount of labor dedicated to programming, scenario development and testing. As AI has moved forward in the development cycle, it has become more firmly integrated as a game engine subsystem that can be modified and refined without upending the entire production workflow.

Rule-based AI driven by finite state machines (FSMs) is commonly used to define the flow of action in combat games, where non-player characters must choose between multiple states as they decide how to act (attack or defend) in response to any number of environmental factors; these behaviors are in turn linked to a number of character animations. Rule-based AI systems are deterministic, defined by complementary sets of pre-determined rules and outcomes, and written through human coding rather than machine learning. The number of possible game states grows exponentially in online multiplayer and open-world games, and places higher demands on these simple intelligence systems.

Game AI is also attached to a number of other code frameworks. For example, the navigation mesh is a data structure used in 3D games that provides the foundation for AI characters to move around an environment; and FSMs may be substituted by behavior trees, another design system for handling character behavior and simulating simple sequential logic. Game AI research has followed a number of interrelated industry relevant pathways. In their panoramic view of the field, Georgios Yannakakis and Julian Togelius (2015) differentiate pre-game or developmental uses of AI (AI in game design and production, in procedural content generation and in setting the conditions for interactive storytelling and social simulations) from game and post-game uses of AI. Game-based AI is commonly associated with the broad spectrum of real-time NPC behaviors (learning, acting and pathfinding), while post-game AI is operationalized during gameplay but understood more properly through the artifacts of play and the data-mining algorithms that produce quantitative information on playability, performance and player behavior. Yannakakis and Togelius consider the distinct operational grounds for game AI, distinguishing between the computer and its embedded AI methods and the end user, as game AI systems are dependent on, that is to say that they can only be revealed and recognized by, human–computer interaction (Yannakakis and Togelius, 2015). Game AI is articulated as a system and a set of methods, and is inherently bound to algorithmic processes and outcomes within a specified context or field of play. The outcomes of game AI are not simply bound to the material or narrative properties of play (what happens during the course of a game); game AI also stimulates a number of player behaviors that include predictive spatial thinking, as players navigate the visual field and determine how to overcome the opposing forces that block their progress. Because AI is embedded in a built information environment and enveloped by a task-oriented system, interface, character set, device or machine, we do not commonly associate what we are being tasked to do with AI *per se*. Yet we do take note of playable media systems that operate at the extremes of predictable or (conversely) unruly AI; these systems tend to fail as natural extensions of ourselves.

Game systems may be one of the more obvious forms of playable media, and indeed, their AI subsystems are part of the gamification of other mediated experiences and the build of game-like experiences and reward systems into quotidian rituals, workaday experiences and other forms of play. While gamified experiences and operations carry with them the signature traces and ritual engagements of video games, they are not always supported by the same technological infrastructures or design frameworks, although they are commonly built with similar data-driven dependencies and intelligence frameworks; and they carry with them the same broad cultural attitudes toward play. Unfortunately, outside of playable media, the horizontal movement of algorithmic intelligence into distinct vertically integrated market sectors often produces only crude forms of gamification—as a number of industries appropriate and capitalize on the incidental properties, rewards and behaviors of gameplay as a means of deepening consumer engagement (Freedman, 2020).

Video games are predominantly conceptualized and designed as environments to enact player agency, and they provide a fertile ground for considering grander ecologies of artificial intelligence, as sites that use AI to purposefully and strategically regulate agency. They also provide a foundation for considering new ecologies of AI, to look outward to see where and how intelligence happens. What are its devices, frameworks and contexts? How is it housed? Where, physically, is it activated, and within what acts of labor or leisure? Video games are themselves not a uniform field. They provide a number of environments for playing with intelligence; and the historical development of video games allows us to see how precisely those environments become more varied and complex, realized in new machines, with new techniques, played within new geographies while themselves creating new spaces of play and colonizing those that have already existed. Both video games and AI have progressively enveloped the outside world, folding it into their operations, whether modeling from it, adapting to it or actuating it (Chang, 2019); and as part of expanding the field of play, both video games and AI have changed the nature of human agency, realizing it "as a manifestation of software, hardware, and infrastructural processes" (Chang, 2019, 12), and shaped the physical and emotional connections (how we formally interact, and how we feel about those interactions) between humans and nonhumans. AI has not yet taken a post-humanist turn; rather, there is a certain duality in how AI is being developed and applied that has created anxiety about the speculative nature of humanness (Luciano and Chen, 2015). AI is commonly deployed in video games as an opposing force; even when it is assistive, it retains its otherness and reaffirms the centrality of player agency. Game AI reminds us that nothing happens without the player closing the informatic loop; algorithms are responsive and dependent on our successive inputs to determine what to generate next. We imagine ourselves playing against or with AI.

Game AI is often distinguished between other uses of AI in that it simply needs to present the illusion of intelligence; the consequences for not truly learning and for simply simulating believable behavior are less disastrous in game-based properties than they are in other industries such as manufacturing. From the broadest perspective, game AI understands the game world as an information space that can be acted upon. The AI subsystem receives information from that world, largely the result of player interactions, and makes decisions about which processes to execute and which animations should be performed.

Two of the most often cited games in histories of game AI are the real-time open-world strategy adventure *Black and White* developed by Lionhead Studios and released in 2001 and *F.E.A.R.* a first-person shooter developed by Monolith Productions and released in 2005. The first of these uses a combination of neural networks and decision trees to simulate learning and decision-making processes, and features an adaptive AI creature that learns from and reflects the player's actions or the actions of the game's villagers, all of which can advance the creature's skills and shape its moral alignment. Richard Evans, the AI design lead on *Black and White*, describes the game's architecture as one that features "representational promiscuity" (Evans, 2002, 567), as it draws from multiple representational models to build a more robust cognitive agent. The creature is built from a belief–desire–intention (BDI) framework that structures its mental state and plays an important role in determining its goal-oriented behaviors, and these elements of belief, desire and intention are strung together from a variety of representations: as attribute–value pairs, perceptrons (algorithms for supervised learning that can weigh their inputs) and decision trees. Evans explains the distinction: "There is something intuitively natural about this division of representations: beliefs and intentions are hard symbolic structures, whereas desires are fuzzy and soft" (570). The perceptron, a form of neural network, invites such a soft and fuzzy logic of desire. What the model illustrates is the ability to develop more advanced learning agents by building a data architecture that accounts for the interactions between the agent and the player and the multi-factored nature of decision-making.

After Lionhead Studios wrapped its work on *Black and White* and was acquired by Microsoft Game Studios in 2006 to create content for the Xbox 360, the studio began work on a project for the Kinect, a console add-on that features an RGB camera, microphone array and depth-sensing and pattern recognition technologies. The technology demo, revealed to the public as Project Natal at the 2009 Electronic Entertainment Expo (E3), was designed to illustrate advanced work with emotive AI and features Milo, a 3D-animated child and Kate, a human player. In its intended form, the demo allows players to interact with and talk to Milo, who can recognize, remember and learn from each conversation and partner with the player on several interactive games. While the project was ultimately abandoned by Microsoft as the company pivoted with the Kinect, its position in the lineage of experiments with emotional and behavioral intelligence reveals the occasional slippage between deeper learning frameworks with what is simply signal tracking, voice (and tone) and facial recognition; in this case, the system listens to and maps the movements of the person interacting with the device, and Milo responds in turn by pulling the appropriate animations from an expandable but dictionary-like database that serves as a surrogate mind. More critically, the project reminds us that the visual and performative contexts of AI matter and indeed inform the beliefs attached to its development and reception; as well, Project Natal represents a distinct effort to reduce the number of mediating layers between the player and the AI agent. By reading gestures, the system eliminates the need for handheld controller inputs.

F.E.A.R. (First Encounter Assault Recon), another significant title in historical accounts of game AI, uses FSMs and a general search algorithm (A*) to control NPC behaviors and to allow them to plan and navigate paths through the game world. Describing the *F.E.A.R.* experience, former Monolith developer Jeff Orkin (2006) notes, "As much as we like to pat ourselves on the back, and talk about how smart our AI are, the reality is that all AI ever do is move around and play animations!" The primary advancement of Monolith's game is the integration of a planning system structured around goals and actions that enables the AI to dynamically solve problems, perform tactical behaviors and use the environment to combat the player. The planning system was Monolith's attempt to manage the game's behavioral complexity and allow NPCs to execute interdependent squad activities in richly detailed game environments; the system simplifies the game's state work by telling the AI what actions are available to satisfy their goals and letting them decide on their own how to sequence those actions (Orkin, 2006). We can see that the push toward environmental realism (in animation, graphic effects and real-time lighting and physics) in *F.E.A.R.* and similarly detailed games continuously pushes AI engineers to work in tandem with game designers to develop realistically embodied character behaviors that can be appropriately animated.

The *Half-Life* series developed by Valve makes a similar use of tactical AI systems. NPC enemies in *Half-Life 2* (released in 2004) run through a number of different decisions when attacking a player and can choose from a number of re-combinatory actions; they can throw grenades, use suppressive fire, and locate and move to cover, making decisions in the context of an active battle (Johnson, 2014). The result, as with Monolith's system, is an AI framework that models deliberative military strategies and employs the best tactics to achieve the collective goals of its NPCs. The NPCs of *Half-Life 2* can act as a group, as they assess the player's location and choose discrete paths to swarm, flank and attack. Having considered the game's combat system, it is important to note that the AI architecture of first-person shooters is, as Paul Tozour (2002b) suggests, multilayered, integrating the more overt intelligence of combat with underlying AI-driven movement, animation and behavior systems; moreover, these systems are active within the localized context of level design, where they must reason through the spatial data of the game world.

The *Left 4 Dead* franchise, a series of cooperative survival horror games also developed by Valve and released in 2008 and 2009, is notable for the introduction of an AI director that can procedurally populate the game environment with enemies and resources and modulate the number of challenges or threats throughout the game (Booth, 2009). The AI director function reminds us that the application of AI in video games extends beyond the control of NPCs. The director function detaches AI code from an avatar or series of avatars and sets it to work on the game world. In this capacity, the AI director still fulfills one of the primary functions of AI systems, which remains resource management; in this case, AI manages the procedural generation of the game world, as well as its objects and avatars. As with other systems, director systems need to be intelligent and they need to act rationally; they are charged with assessing the entirety of the game world, with continuously assessing its current state, shaping its dramatic unfolding, changing the world to fit the context of play, and measuring and managing its pace (Thompson, 2014). *Alien: Isolation*, a 2014 first-person stealth game developed by Creative Assembly (published by Sega through a longstanding franchise license with 20th Century Fox), follows a similar dyadic AI model that pairs character AI with director AI and manages the task orientation of specific behaviors while also independently monitoring the state of play. The central gameplay of *Alien: Isolation* is controlled by two distinct AI systems. The activities of the alien xenomorph (the game's antagonist) are defined by a behavior tree architecture that subdivides its actions into searching, hunting and investigating as it pursues the player; while a second system, the AI director, tracks the locations of both the xenomorph and the player, and uses this knowledge to manage the pace of the game (Thompson, 2020b). The interactions of these two distinct systems, one character-based and limited in perspective, and the other developer-based and omniscient, complicates the player's strategy and heightens the game's sense of menace, as the directorial point of view is not represented in the visual field and remains largely indecipherable. These two systems highlight how intelligence can hold multiple, interdependent and expressive gameplay functions to create an operational whole.

Cyberpunked: The Failure of Environmental Intelligence

Even when it is embodied, AI articulates more than character behaviors; it functions as an organizing principle for both the game world at large and as part of functional level design. This may be most apparent when it is used to manage the resources and activities of virtualized large-scale outdoor urban environments; these spaces can be crowded with pedestrians and motorists, each with a unique agenda and set of behaviors, which makes them a challenge to simulate and articulate as an organic and logical whole.

In October 2020, Take-Two Interactive Software filed a United States patent application for a "System and Method for Virtual Navigation in a Gaming Environment," offering a glimpse at how to manage and simulate the unpredictable yet patterned nature of large-scale outdoor environmental mechanics. The patent outlines the operations of traditional navigation systems before it proceeds to detail specific performance improvements for managing objects in a multiplayer network:

> Since currently-available multiplayer gaming systems are deficient because they cannot provide realistic movements for non-player objects in a virtual world without increasing computational resources and/or restricting game development/design, a system for managing nodes and node graphs relating to non-player characters that provides virtual navigation and management can prove desirable and provide a basis for a wide range of network applications, such as creating a realistic virtual world that is not limited by hardware and software limitations.
>
> *(Take-Two Interactive Software, 2020)*

While no particular game title is mentioned in the patent application, and the filing seems to simply be pointing to the value of moving event logistics to a cloud-based system that can handle the bulk of the in-game pathfinding or navigational work, Take-Two Interactive Software is best known for the *Grand Theft Auto* and *Red Dead Redemption* franchises, two highly successful open-world game series published by its subsidiary Rockstar Games. The proprietary Rockstar Advanced Game Engine (RAGE), which has been used across both franchises, manages the behaviors of non-player characters and the associated event handling mechanisms of each game world, and is able to balance the directives of exploration and campaign-based play and the different demands placed on defining movement and combat.

Open-world games tend to balance explorative and goal-oriented work by including various signposts of progression that include a variety of forms of "leveling up" (acquiring new skills or accessing upgrades) that often serve as keys to unlock new areas of exploration. Still, these previously unseen maps are grown from a cohesive base design of the game world at large and align with a set of overarching organizational principles. As we proceed through an urban open-world game, we are commonly invited to play in and with the city's properties and resources (Freedman, 2020). The city is tied to an engine that renders it visible and allows the player to draft an itinerary; and AI guarantees continuity in the game state while also managing the complex tasks, interactions and atmospheric conditions of the larger ecosystem. Much like its counterpart, the AI used to manage smart city systems (for traffic management, public transportation, parking, surveillance, waste management and so forth), environmentally bound game-based AI needs to attend to both discrete and integrated activities and perceive and regulate these as part of an interdependent supersystem. AI, particularly in a procedurally generated world, needs to be attentive to what the player is doing and where the player is going, as it stitches the environment, its occupants and its systems together. Players tend to notice when the world falls apart.

On December 14, 2020, CD Projekt Red issued an apology through its Twitter account. The developer was forced to address the poor technical performance of its latest release, *Cyberpunk 2077*. Suggesting that it would fix multiple bugs and crashes through a number of software updates, the developer assured players that it would address the most prominent problems that were plaguing last-generation consoles (the PlayStation 4 and Xbox One), but also made it clear that full refunds would be available through Sony and Microsoft, as Sony began to pull the title from its online store. What players experienced as uncharacteristically annoying NPC behavior, misbehaving assets, poor driving mechanics and erratic gameplay elements pointed to a host of underlying technical flaws in the game's AI and physics systems. Since the game's release, players have posted videos of their experiences with vehicles driving through other vehicles, violating all material laws; similar demonstrations of the improper physics and surface geometries of trees have appeared in several sub-Reddit forums. Other players have recorded their experiences with repeated bugs, crashes and visual glitches, editing these together in lengthy evidence reels. Most customers have felt betrayed by a game that lingered in development, suffered from repeated delays, was rushed to completion and ultimately released in a somewhat unfinished state, after being repeatedly promoted as the next revolution in open-world gaming. The cinematic trailer revealed at E3 in 2018, marked as game engine footage, features a gritty and violent urban metropolis that promises to immerse the player in a richly defined and densely active city of the future. Rather suspiciously, as the game's launch date approached early reviewers reported receiving PC codes to review the game while being denied copies for the PlayStation or Xbox. Many of the reviews that began to surface near the time of the game's release in December 2020 were based on PC playthroughs, and when reviewers were finally able to get their hands on console copies, they reported issues with the game freezing or locking their systems, slow loading times, glitches with asset geometries and textures, periodic drops in frame rates and breaks in the enemy AI. For many reviewers, the console game was unrecognizable from its PC counterpart. What these early reviews highlight are the myriad interdependencies between software and hardware, and between game studios and console developers; as CD Projekt evolved its proprietary game engine it failed to fully optimize its game to meet the demands of disparate platforms and consider how its programming and AI subsystems needed to negotiate the distinct runtime environments of last-generation and next-generation consoles.

In 2016, the CD Projekt Capital Group received more than $7 million from the Polish government's GameINN program (which issues grants financed by the National Center for Research and Development in the video games sector) to further its research on multiplayer city creation. The company submitted and received funding for four interrelated game development projects: "City Creation, Seamless Multiplayer, Cinematic Feel, and Animation Excellence" (NCBR, 2016). In March 2017, CD Projekt filed a request for proposals with the Polish government to subcontract the services of an engine programmer for its City Creation technology: "City Creation—Comprehensive technology for the creation of 'living,' large scale urban environments playable in real time, based on rules, artificial intelligence and automation—including innovative pipelines, data layout and supporting tools aimed at complex, high-end open-world games" (CD Projekt, 2017). The City Creation technology proved to be one of the biggest problems with the finished *Cyberpunk 2077* game. The AI of the inhabitants of the game's densely populated Night City was demonstrably unruly (at best) at the time of the game's release; early players recorded their experiences with NPCs randomly disappearing, cars driving through obstacles, and police shooting and killing their unarmed avatars. While the studio has suggested that urban design experts were consulted on the build of the game's featured Night City to assist with laying out its infrastructure and its architectural footprint, the failure of the game can be measured against the promise of advanced crowd and community systems, and the collapse of the physics and logic of the game's open world. For a game that folded artificial intelligence into both plot and story (and even gameplay, as players can modify the gender-fluid protagonist V with cybernetic limbs to match their play styles), the erratic performance of the gameplay system is particularly ironic. Of course, the broader problem with the game's faulty AI is its failure to support an immersive storytelling experience despite the research that informed the futuristic design of the game world. In this case, game art and game programming failed to align. During a May 2021 livestream on Twitch, Pawel Sasko, one of the game's lead quest designers, attributed the faulty NPC AI to design intention, to purposefully making the AI behave like a person with somewhat faulty judgment. Gamers on Reddit quickly took issue with Sasko's explanation; and a patch released in February 2022 that introduced new downloadable content, new character customizations, and a number of AI fixes to NPC combat and crowd behaviors seems to belie the designer's earlier appeal. After CD Projekt's success with its previous title *The Witcher 3: Wild Hunt* (released in 2015), the failure of *Cyberpunk 2077* may seem surprising; the earlier game was also buggy, but the richly detailed world and its layered storytelling (which drew from well-regarded source material) allowed players to forgive the game's occasional performance issues. Both *The Witcher 3* and *Cyberpunk 2077* are built from iterations of the studio's proprietary REDengine (versions 3 and 4, respectively), and feature vast open worlds that can accommodate complex, multi-threaded (nonlinear) storytelling; yet the City Creation subsystem was a new addition for the fourth-generation engine, and the breakage seems to be fairly sweeping in the latter game's AI, having an adverse effect across rendering, physics and NPC behaviors.

Cyberpunk 2077 is a notable case of the failure of AI largely because it reveals much about game labor, studio workflows and the pressure to market what was a flawed product; yet this particular studio is not alone in failing to meet pre-release hype and expectations. The game also reveals much about the troubling slippage between background and foreground AI (Treanor et al., 2015), when systems that are designed to support gameplay actually impede it and break through to the surface. Throughout its lifecycle, *The Sims*, a series of worldbuilding life simulation games published by Electronic Arts (that have continued to evolve through the 2014 release of *The Sims 4)*, has had uneven AI performance. This has been quite noticeable in a game where the AI can take over and control the actions of its in-game avatars. *The Sims* AI works within a hierarchy of needs; each Sim can consider its internal state and determine how to act to fulfill its own needs. As the game is designed with an object-oriented operating system, Sims can adapt to any new object that is dropped into their world and determine how it might meet their existing needs. The publicly shared list of cheat codes that has circulated widely online suggests a number of players have experienced points of failure in the game, and have found ways to negotiate their issues with the game's intelligence framework or have simply hacked it to match their own play styles; among them, *The Sims 4* command line, "fillmotive motive_[motive]" can resolve a Sim's bladder, energy, fun, hunger, hygiene or social need. These social posts suggest players respond to game AI in unique and varied ways, and that failure can generate productive, collaborative problem-solving. In a series with such a studied history of self-directed worldbuilding, and in which AI is purposefully designed to be managed, *Sims* players have become quite dexterous and forgiving in their responses to failure, and have embraced the game's do-it-yourself ethos by complementing its AI systems with additional code work. In contrast, CD Projekt Red violated the trust of its fan base, and the negative publicity surrounding the game was only heightened by public statements from studio employees about longstanding unfair labor practices.

CD Projekt Red came under fire in 2020 after the studio told its developers they would have to work mandatory overtime to meet a November release date, pivoting from a statement made the previous year that suggested overtime would be non-obligatory. Similar crunch policies (of compulsory and often unpaid overtime) have plagued the video game industry for years, with a number of other studios including Electronic Arts, Rockstar Games and Naughty Dog being accused of equally exploitative labor practices. But *Cyberpunk 2077* had been hyped for almost a decade. The studio first announced the title in 2012 when it held special promise for next-generation consoles, and it re-announced the title again in 2018 only to showcase the game the following year with a clip featuring actor Keanu Reeves (whose likeness is included in the game as the playable character Johnny Silverhand). Several additional delays were announced throughout 2020, even after the game had gone gold (meaning it was presumably ready or near ready for publication) in early October. But gameplay footage had still not appeared.

One of the elements hyped during the game's pre-release was its character editor, a key feature of a cyberpunk fantasy that treads between bodies and machines, but as many players have noted, the promise of a trans-humanist future was under-realized in customizations that could only produce a number of static choices rather than representing the body as an infinitely malleable canvas. Yet these surface concerns which provide the player with a significant amount of control over the appearance of the game's protagonist (including customizable genitals) do not speak to the inordinate number of performance issues they experienced while actually playing the game. As I have suggested, game art and game design were undercut by game programming; the game's intelligence systems ultimately interfered with its (visual) runtime grammar, as these systems were never fully debugged despite the game's repeated delays. At the time of its release, the glitches in *Cyberpunk 2077* produced unexplainable visuals and shoddy performance; the AI seemed indeterministic. Customers deemed the game almost unplayable; many reported full console crashes and most demanded refunds. Players openly shared their frustrations with the game on social media, posting videos of their gameplay issues.

AI does not always succeed in video games; the most common glitches impact AI-based companions, enemies and other NPCs, and that impact is multiplied in crowds; these characters are all running a number of checks, using finite state machines, behaviors trees, utility AI or goal-oriented action planning to identify the current state of the world and decide how to act within it as they pursue a certain goal. The faults of AI are lodged in its code and can be found in any one of its associated systems but are most often connected to decision-making and action sequencing. These glitches are commonly called out by players because they have a significant material consequence.

The problems with AI in *Cyberpunk 2077* remind us that data systems are connected to and influenced by a number of physical constraints and technical conditions, and that AI is shaped by the informational character of its medium, the labor that supports its authoring and integration, the development of its (network) infrastructure, and its relationship to our bodies—where playability matters. While we might associate crunch time with putting the finishing touches on a soon-to-be released game, the failures of *Cyberpunk 2077* and the facts of its production reveal the deeper dependencies on AI programming that exist throughout the game development cycle, where AI is not simply a fully realized front-end subsystem, but a deterministic element whose programming needs to be re-negotiated throughout a game's build. The many failures of *Cyberpunk 2077*, both as a playable game and as a published object, bring our attention back to the platform and highlight the importance of the labor associated with building AI. To fully acknowledge the labor invested in manufacturing AI systems, we need to disentangle the way AI is made from how it is used (Moore and Woodcock, 2021). We need to understand how AI works in practice and how that practice is integrated into related systems.

Beyond Gamification

Modern game AI plays a much wider role in game development and runtime execution than building NPC behaviors; and its work in automation, co-creation, procedural generation, runtime optimization and personalization accounts for its myriad attachments to other media and non-media industries. The movement of AI across these multiple forms and functions should prompt us to examine what development studios are putting into the back ends of their systems, where AI can automate analytics and harness consumer meta-data and help studios re-calibrate their investments. Machine learning and behavioral recognition suggest in very real ways that video games can understand what players are doing and provide mechanisms for developers to build relationships with their consumers. Video game technologies have extended their algorithmic influence into the machinery of the information economy (Finn, 2017), bringing with them the humanistic attachments of play-centric algorithms that make software so welcome and familiar.

To understand the influence of the video game industry on the culture writ large, we need to continue to move beyond the fundamental concept of gamification, which speaks largely about game design elements being used in non-game contexts (Deterding et al., 2011). We need to maintain a systems approach that understands the video game industry as a series of interlocked intellectual properties that include game titles and software engines. The video game industry has developed a generative momentum for playful design that is grounded in its software architectures and services. Gamified experiences and operations carry with them the signature traces and the ritual engagements of video games, but the more important work being performed within them is happening at a systems level as they harness consumer data and intentions.

In her deeply critical examination of gamification, Sonia Fizek (2014) distinguishes between playfulness and playability; the former, Fizek suggests, reflects an attitude or an approach to engagement that may be supported by but is not bound to one particular design scheme, while the latter more aptly describes what have emerged as dominant forms and mechanics of engagement. The significance here is that the distinct pursuits of being playful (a standard of engagement) and playable (a measure of technique) inform how game technologies are mapped onto other media forms; the engine-based systems and subsystems of game design and development have already formed deep roots in the broader field of information design, and are shaping how we see, understand and engage with the world (Zimmerman, 2009).

If we are to move beyond industry convention and create more expansive forms of gameplay, we need to acknowledge the contexts in which games are played and the computational models that are being built to bring them to life and, in turn, organize player behavior and set perspective. These computational models are often used to inform procedural content generation and real-time visualization in other media forms, porting the tenets and practices of game design to new initiatives. Can game AI methods be extracted for real-world uses? Research on AI game-playing methods that leverage reinforcement learning to acquire domain-specific knowledge and solve domain-specific tasks can be used to test for weaknesses in a broad range of critical systems; and by using game-playing agents to simulate complex real-world phenomena, we might accelerate our response to the climate crisis and any number of social, economic and environmental challenges.

If we hold onto the dual concepts of playability and playfulness, we will be better equipped to discuss both the "how" and the "why" of playable media. As past research had prophesized and current work is now bearing out, AI is a generalizable technology; and its inherent openness makes it available to cross-industrial applications. These existing and emerging applications are being supported by private equity and venture capital as well as research funded by global technology giants such as Amazon, Apple, Google, Meta and Microsoft; and while these investments have established new pathways for artificial intelligence, they have also created an uneasy alliance between the economic gains and social benefits of AI-related activities (Prug and Bilić, 2021). Our culture is being fundamentally transformed by playable media platforms, and while we can readily evaluate their quality and measure their impact, we need to do a bit more digging to understand the influence of the software and intelligence systems that run them.

3

CHEATING DEATH

Artificial Intelligence, Non-Player Characters and the Logic of Pandemic Culture

The Last of Us Part II, the second installment in developer Naughty Dog's video game franchise, was released in June 2020 at a moment when Coronavirus cases were still trending upward around the world, with notable flashpoints in the global south and in several regions of the United States, and the true scale of the pandemic had already been well-documented in epidemiological reports from the World Health Organization. National governments moved at different speeds to introduce emergency measures to stop the spread of the disease, shore up healthcare systems and head off economic collapse, while citizens responded far from uniformly to state policy and action, and with differing degrees of trust in experts and authorities, often reading scientific advice and public health advocacy through a political lens, even as shared geography should have led to common and collective action.

While the parallels between the viral outbreak in *The Last of Us Part II* and the outbreak in the world-at-large are all too self-evident, with the former set in a post-apocalyptic moment that from our current vantage point may seem like a fantastical projection, the less evident tether that connects the two worlds of contagion (one a hermetic narrative, and the other still unfolding) is their mutual dependency on what Mika Aaltola (2012) refers to as "politosomatics," a term he deploys to describe how the anxieties of each individual are inherently measured against and bound to a global collective. Pandemics trigger a heightened embodied response; people translate their fear and risk into material practices (hoarding, cleaning, avoiding public spaces) and ritualistic containment strategies. While these mitigation measures may be grounded in public health policy, they reify the existing imbalances of wealth and power that circumscribe public health; and though they have a communal orientation, they are operationalized at a hyper-local level and necessarily focused on individual wellness. It is in the context of

DOI: 10.4324/9781003225072-3

evolving and uncontained pandemic threats that Aaltola suggests, "people often sense their wider surroundings and their world as embodiments with which they identify in varying ways and on behalf of which they worry" (2012, 2). Individuals construct their own disease behaviors from this ground, as they attempt to make sense of uncertain levels of risk and manage the larger socioeconomic conditions that exceed their control.

The Last of Us (released in 2013) and its sequel remind us that the rules for navigating space during a pandemic require a dynamic assessment of risk—rules that depend on data and are embedded in independently authored systems that meter tactical thinking and freedom. To fully account for the interplay of artificial intelligence, visuality and agency, we need to move toward a common framework for understanding how agency operates through computer programming and algorithmic projection and hold onto both data and image in our examinations of playable media, where artificial intelligence is a functional property of the modular systems that undergird diverse object-oriented contexts. Video game spaces, their environments, characters and mechanics are embodied practices, actualized and continuously negotiated by players; in survival games, the components of game space are balanced by design to regulate the player's power and vulnerability, and to encourage the player to remain ever vigilant, to occupy and monitor any number of "differentially exposed bodies" (Aaltola, 2012, 1). The balance of power in video game spaces, which tilts in favor of the developer, may not be political in nature, but the experience of duress is a complicated affair, produced by an algorithmically organized world order, a code space that uniformly governs bodies and their wider surroundings. Following a number of strategic design and narrative principles that are visual but grounded in computation, and that shape and gate information, the development team translates the environment into a fear-producing mechanism.

The Artificially Intelligent Body: Reading Non-Player Characters

Video game play rules and structures are made possible through artificial intelligence frameworks—frameworks that also structure non-player bodies that must be read and interpreted for players to act, survive and progress. Non-player bodies are designed to mediate player agency, and that design intention is understood by players; non-player bodies influence player interaction and movement. Moreover, non-player bodies are the primary drivers of physical panic in survival games. Their movements are tied to what is experienced as an uncertain algorithmic indeterminacy, even as they follow strategic design principles; although the order of the game world is consciously structured by programmers and developers, it is only partially revealed to players. The limited circulation of power and control through the careful gating of information produces the primary experience of stress and fear during gameplay. Within its narrative framework, a game's information architecture is continuously translated into distinct animations, environmental transformations and object-oriented movements and displacements; together, data and visuality contour player agency; although artificial intelligence is a game design subsystem, it has a deterministic influence on the visual field.

Janet Murray's (1997) discussion of agency has informed more contemporary critical considerations of play and power in video games. Murray attaches agency to "the satisfying power to take meaningful action and see the results of our decisions and choices" (126). The sense of power and control that video games afford is attached to their status as computational systems that call for input. The player acts, the engine and its AI systems react in an ongoing cycle. The player and the game text (its coded architecture) are active and reactive, while the game developer and in particular the game programmer creates this circuit of relations and effectively authors the environment. One of the difficulties in integrative media studies is giving balanced attention to the player-consumer and the developer-producer, an act that requires more studied attention to the informatic (computational) systems that drive visual culture and acknowledges the latent pleasures of code for those who author it and those who read and play through it.

My approach to video game analysis is aligned with the game development process and a division of labor that draws a critical and practical line between the immaterial artificial intelligence frameworks that drive non-player characters (NPCs) in video game space and their materially bounded bodies—distinctions between game programming, game art and game design that are nevertheless interwoven within the production pipeline and in the bounded intellectual property (the playable game). In sheer numbers, non-player bodies dominate game space, and while they are often understood in studies of visual culture as repeatable and often stereotypical archetypes, they are complexly layered and communicative software systems. Non-player characters exist as medial figures; to fully imagine them, we need to consider their external forms, their narrative functions and their status as data containers (Schröter and Thon, 2014). To develop a fuller understanding of the visual body politics of video game representation, we need to take note of how artificial intelligence, as a part of core software architecture, generates affective behavior. By studying the performance of loops, cycles, boundaries and vectors (the systems that govern non-player bodies) and understanding these relations, players are able to gain control over playable space; they are able to "cheat" a system using scene analysis to translate a general strategy gleaned from complex operational patterns into sequential tactics. The player's decisions are driven by certain overarching assumptions about game logic. To survive, to progress, is to know non-player enemies (in the case of *The Last of Us* and its sequel, categories of infected bodies that include Runners, Stalkers, Clickers, Shamblers and Bloaters) and companions as communication systems and data systems, to read embedded algorithms as overdetermined pattern agents in a series of antagonistic or supportive bodies. Algorithmic knowledge (a facet of procedural literacy) is a fundamental skillset that structures most gameplay experiences. In a parallel consideration of the AI frameworks of *Alien: Isolation* (released in 2014), Jaroslav Švelch (2020) suggests that video game players welcome the unpredictability and challenge presented by autonomous AI agents, but also expect that they will come to understand and master the behaviors of their opponents over the course of play. What *Alien: Isolation* and *The Last of Us* reveal are the interdependencies of multiple AI systems and agents; as I have already discussed, the former pairs character AI with director AI, while in the latter, the player must negotiate both enemy AI and companion AI (Ellie), while both enemy AI and companion AI must understand what the player is doing. *The Last of Us Part II* expands the roster of companion characters from the first game to match the story arc, but the dynamics between AI systems remains intact. Enemy and companion AI systems are drawing from a unified data set that includes localized knowledge about the player, but they are designed with different priorities, actions and animations as they navigate and are guided through the game world; and to progress, the player must be correctly positioned within this dyadic system.

The desire to situate AI as an expressive form that links gameplay, game production and game studies emerges from two convergent pathways: design practices that seek to simulate intelligence and playable activities that look for signs of life in readable game behaviors (Mateas, 2003a). AI frameworks are used as assistive building tools to accelerate game development workflows, automate and enhance the design process, create and populate complex virtual worlds and guide the production of play; and they are used as behavioral tools to guide the experience of play and test and analyze playability. The practical design and implementation of video game AI are aligned with its generally desired executable expression, which is to produce intelligent behavior that is believable and balanced in its predictable and unpredictable nature; that balance creates the challenge found in gameplay. What emerges is a language of emotive systems that exceeds the purely technical conceits of AI. Michael Mateas (2003a) suggests, "for game analysis, game AI is the language of intentional behavior."

To understand how space is designed to create or delimit possibilities, we need to unpack its technical architecture; game space may be occupied and activated by bodies that seem to have purpose, but it is ultimately governed by AI, a system that is designed and bound to several other operating tools and layers: among them, the interface and the controller. At the same time, we need to move beyond understanding the structures and (mathematical) rules of games; within game space, a phrase that aptly describes a video game's governing technical architecture, "play space" emerges as a more dynamic act of being in the world in relation to other entities (Möring, 2019, 232). The dimensions of play space, the sequential actualization of game space, are in constant flux (although bounded by the game world) and redrawn with each playful encounter with intelligence. While game researchers Petri Lankoski and Staffan Björk (2007) suggest that fully believable NPC behavior is an impossible goal, they also propose believability should be reframed as a consideration of more purposefully designed gameplay that situates behavior in its appropriate environmental context; to this end, AI has widened the scope of character design beyond responsive context-dependent character dialogue to create more resonant systems that impact the style of play and the decisions and movements of the player. The development of game AI follows some of the generalized design rules that always connect form to function, as the runtime execution of a game's underlying code always governs the range of abilities of its characters, and those abilities are in turn attuned to the design of the game world (Nitsche, 2008). AI may be formally embodied by an NPC, but as an executable function, it is also attached to and determined by its environment.

Pandemic culture propagates a similar world view of calculated environmental unfolding, and is a particularly poignant narrative video game trope, as it requires us to read and navigate the bodies of non-playable others in a dynamic landscape of limited resources and dotted with viral threats that challenge our cultural memory and set the conditions for new physical affordances. Non-player bodies are under-analyzed forms, often regarded as assets or game objects rather than complex performative character systems with calculated volatility that can teach us about public sector mobility. By playing and reading through *The Last of Us* and its sequel, as well as parallel survival narratives, we can reclaim or at least develop a deeper understanding of the role of non-player bodies that seem to exist purely for their technological necessity and playful complexity. These bodies are defined by complementary processes of containment that matter in the game production pipeline: the procedural and economic imperatives of software engineering, the design and functional imperatives that attach a continuum of monstrous others to a common codebase, the engine-based algorithms that regulate their (classified) behaviors, and the geometries of 3D space that define level design and the viewable field.

Game AI as a Technical and Cultural Form

There is a considerable interdependency between the broad cultural work of artificial intelligence as an instrument of crisis management and the role of artificial intelligence in game-based pandemic narratives and as an infrastructure of visuality, though crisis and play are articulated through distinct technical structures; they use different forms of AI, with the former driven by the methodologies of data science and the latter by the execution of real-time expert systems common to video games. Even so, artificial intelligence is one of many mediating forces that define our response to crisis. AI is not an independent agent; it only acts in concert with other systems, as part of a platform that may involve a range of actors (clients, consumers, project developers, regulators and so forth) and where it is tasked with doing something. AI is meaningful only in practice, and in the context of pandemic response, its role is complicated by established disease rhetoric. Aaltola notes, "disease, as a socially interpreted physical and psychological process, is fundamentally shaped by prevailing political culture and practices"

(2012, 3). Pandemic game narratives are operationalized in the context of culturally validated disease experiences, and their underlying systems are designed to strengthen the associated imaginaries of fear and risk that accompany societal collapse. *Plague Inc.*, a real-time strategy game developed by Ndemic Creations and released in 2012, invites players to create and evolve a pathogen in an effort to infect the entire human population. The mobile game uses an epidemiological model with a complex set of variables to simulate the propagation of a player-produced disease and track the evolving pathogen as it adapts to various environments. The objectives of the game include killing the world's population with a series of unlockable pathogens (that include bacteria, virus, fungus and parasite) or enslaving the population by formulating unique mind-controlling plagues. Simulators like *Plague Inc.* use real-world logic and embed it in synthetic scenarios, but also draw from existing disease media for their narrative fascination and affect (Mitchell and Hamilton, 2018). The 2014 Ebola outbreak in West Africa led to a dramatic spike in *Plague Inc.* downloads. The game's timely rise in popularity accompanied a growing intellectual curiosity with understanding how infectious diseases work, a curiosity that is often awakened during times of pandemic scares. To that worthy end, despite its morbid(ity) goals, *Plague Inc.* serves as a primer on basic disease reproduction numbers and the biological, socio-behavioral and environmental factors that govern the transmission of pathogens.

Deep in their code base, survival narratives are scripted to enact and display bodily discomfort. In this context, AI is meant to induce harm, and game programming, art and design work as hegemonic embodiments of visible stress in the world at large; and the serialized intellectual property, the game franchise, codifies this vocabulary to effectively govern play. Unraveling this interdependency allows us to lay bare the affective dimensions of AI systems; to do so, we must simultaneously hold onto a holistic view of the game world and tease apart the particularities of its code. Gameplay rules and structures, the evidence-based decisions that impact game outcomes, are made possible through a game engine's underlying artificial intelligence frameworks. While decision-making is a form of ethical ideological engagement, from a purely mechanical perspective, gameplay can be understood as a scripted cycle of interaction that leads from information gathering and analysis to decision-making, interaction and reactive operations and animations (Freedman, 2020).

Game AI draws from the broader field of artificial intelligence and the techniques associated with case-based reasoning, decision-making, inference and cooperation within a multi-agent system (Tozour, 2002a). The common conceptual schism between game AI and AI in its broader industrial application is the relative emphasis on play and imitation, the nature of exactness, and the practical value of demonstrably true intelligence; games commonly associate fidelity with visual pleasure, yet non-game industries associate fidelity with the signs of human cognition and the affordances of natural language processing. In video game performance, flawless intelligence is not always the appropriate goal, as it can be too restrictive and too limiting on an otherwise more dynamic range of play skills, experiences and strategies. In industrial contexts, AI must demonstrate infallible task-oriented intelligence. Yet in both applications, in game and non-game industries, the intelligence of AI systems is always measured against human performance.

While game AI is associated with mechanics, it is a property of the engine, itself the building block of efficient real-time visualization (Freedman, 2020); the game engine is a software architecture that manages an array of computational tasks central to game development and execution. As an algorithm developed by the programmer that can be associated with data management, AI nonetheless creates clear signposts of intentionality and agency. We may consider AI as a component of a character or a game object more generally, in relation to the physics, render and transform components of the same object, but to develop a fuller understanding of a game object, we need to take note of how these components generate affective behavior.

The AI systems of modern games give non-player characters the ability to make decisions about where to move, and to think and act strategically as they do so (Millington and Funge, 2009, 8). NPCs interact with players, the 3D environment and with each other, and their actions in the game are driven, at least in part, by real-time decisions made by the NPC algorithm regarding the current state of the game and the environment. Games often use a network of nodes (or graphs) to provide paths for NPCs to traverse the game world. Each node corresponds to a physical location in the game world, and these can be connected to one or more other nodes in a graph architecture. Typically, each node has associated metadata that provides relevant game data that can be accessed by the NPC to determine its movement. To move around the game, the NPC creates a path that follows a series of nodes that represent a route from a starting location to a destination. AI systems bind NPCs to the governance structures of the game world and the particularities of level design.

The contextual operations of AI across distinct playable and non-playable character bodies are driven by a number of underlying conditions, embedded in the codebase, that, to varying degrees, prompt player identification, call forth distinct actions and, by nature of their animations and vectors, allow for a distinct set of physical transformations. In video game development and execution, artificial intelligence is a gameplay subsystem; it is one part of a governing game engine and operates within the rules that organize the game world. AI commonly functions as a goal-oriented gating mechanism. AI can be used to help NPCs gather path information (pathfinding is then executed in tandem with a physics system that provides collision information around the edges of an area, and around all interactive objects and player characters), imbue an NPC with line-of-sight information and other perceptions, and provide dynamic avoidance within the game world model (Gregory, 2009, 48). The AI subsystem operates in tandem with other game engine subsystems that include the physics engine, which simulates the formal influence of real-world physical systems. The game engine is comprised of a number of interlinked calculating software systems that draw from a shared set of data. Players progress by understanding the underlying structuring tendencies of AI-driven systems and by deciphering the work of other bodies—bodies under the influence of an engine known to the player only through its visual signatures. Of course, in addition to reading situated intelligence, players must also understand the rules behind the game and its control schemes, as well as broader aspects of simulation, physics and the built environment. Non-player characters further complicate the playing field, and are not simply assets or objects; they are mutable containers for the processes of coding and execution that undergird their many forms, and players must learn how they function in each game context. Michael Mateas and Noah Wardrip-Fruin (2009) situate this work, of writing and reading code, as a matter of "operational logics" insofar as the technical strategies of game development are both literal (and functional) in that they underwrite gameplay, and abstract in that they create a set of interpretive outcomes available to both game designers and players; operational logics effectively bind the state of production to the state of play and they link graphical functionality (the matter of game engine architecture—movement in the general sense, and physics and collision detection in the particular) to textual design (the visual matter of game art and assets). These interlinked graphical and textual logics (Wardrip-Fruin, 2005), bound through real-time rendering, are the primary structures of play.

Game AI is more than a mechanic for NPC action; its operations are more expansive, assuming multiple forms and functions within the game world. AI does not simply remain invisible as it monitors and controls the game state. AI is also something to be encountered, it can be observed and interacted with as an alter to the player's ego (Straeubig, 2019). AI can be stood up as an independent agent, given form for the player and bound to a communicative body; and it can monitor the state of the game world and actively shape in-game events. AI can even act as an automated world builder that fosters more passive forms of player engagement (Fizek, 2018). While it is beyond the scope of this analysis, AI has a place outside of traditional gameplay in self-playing games that leverage deep-learning architectures; here, AI acts as an independent "player" or agent, monitoring, shaping and, in some cases, creating the conditions of the world. AI can also be tasked with authoring a procedurally generated universe, continuously iterating and rendering the field of play; and this game work can be extended to autonomous systems that depend on machine learning to monitor the built environment, attend to its physical structures and meet the demands of sociability. As social frameworks, autonomous systems (e.g., self-driving cars) are capable of interacting with other agents, measuring their performance and tailoring their behaviors in reference to external circumstances and across a series of interdependent operations.

As AI entities, non-player characters must demonstrate multiple behavioral properties; they must be autonomous, adaptive, goal-oriented and sociable. NPCs must be able to react and adapt to changing environmental conditions and inter-act as part of a multi-agent application. Yet NPCs develop in rather limited ways that serve the parameters of the game world; while they may learn and adapt from their interactions with players, NPCs can only perform according to their software architectures and the design and narrative principles of the overall game (Yannakakis and Togelius, 2015). They may be autonomous agents, but NPCs need to respond in meaningful ways to the actions of the player (NPC behaviors need to make sense) and within the capacities of the game world and its envi-ronmental design. While video game AI is designed as a form of computational automation, it is done so with the player in mind (Fizek, 2018). Video game AI is purposefully designed as a decentered force that reacts to what the player is doing.

Game AI is a larger field of research and development, and while it is central to NPC behavior learning that reads and models itself after the player, game AI is just as deftly applied to the areas of procedural content generation (the semi-automatic creation of game content, notably levels and maps and during gameplay), computational narrative development, intelligent agents, AI-assisted game design and a number of other areas that influence each other while placing varying emphasis on the computer or human (the end user developer or player) perspective (Yannakakis and Togelius, 2015). The video game industry has largely moved on from solving the NPC problem (how to model human-like behavior) to pursue the more holistic problem of interweaving game design with game technology (Yannakakis and Togelius, 2018); character studies have yielded to player and worldbuilding studies and the production of affective-based procedur-ally generated content. Yet there are lingering performance issues associated with NPC AI and companion AI (the latter are assistive rather than antagonistic NPCs, and are often used for single player partner-based combat and puzzle solving); and the AI problem, however local or global, can only be solved by reading and writing from each small player interaction. AI systems are not simple consumer technologies; they draw from a network of resources, and continuously ingest, analyze and optimize data, as they produce the playable field (Crawford and Joler, 2018). With each interaction, an AI system becomes more aligned with the user's goals and intentions.

A Case Study in Artificial Intelligence: Naughty Dog

Video game developers that have the resources to assemble their own software frameworks approach video game development as an opportunity to actively build new approaches to play and the associated code work of play, and establish unique but decipherable operational vocabularies; for these developers, the game engine is a customized solution to both novel and systemic (well-known) problems. The engine establishes a game's principal machinic affordances, but other operational limits are set at several points in the circuit of production, in environmental and character builds, in the allocation of in-game resources, in the interface and in the control scheme. Those coded connection points between problems and solutions may of course be repeated—they become an organic pattern language, a focused environmental solution and a unique intellectual property when attached to a proprietary engine (Freedman, 2020). Artificial intelligence is one of several engine subsystems that produce a game's recognizable operational logic and provide the player with a pathway through a series of executable processes that equate data with regulated agency (Mateas and Wardrip-Fruin, 2009).

To illustrate the relationship between artificial intelligence and player agency within a focused set of objects, I turn to the AI systems produced by Naughty Dog, a California-based game developer operating as a subsidiary of Sony Computer Entertainment. Naughty Dog has evolved its AI systems across two significant franchises: *Uncharted* and *The Last of Us. Uncharted* is a globe-trotting treasure-hunting adventure series, while *The Last of Us* is a character-driven pandemic narrative that traverses multiple cities across the United States. The games in the two series are formally aligned in their use of geographic specificity and detail, and structurally aligned through their respective AI frameworks and the calculated movements, activities and awareness patterns of their non-player characters; modular skill trees wed these structures to knowable character classes and animation sets and activate the environment in precise ways that match the attributes and abilities of each class. Naughty Dog's purposeful approach to game AI links the work of its programmers and artists within the production pipeline, and functionally makes these coded skills and associated classes knowable to the player. By mastering these AI systems, players are able to understand the technological affordances of the played environment and develop responsive survival strategies. To make headway through each game, players need to negotiate bodies and environments, and need to consider each from a tactical perspective. Conflict is oftentimes unavoidable; players are not always given an opportunity to decide whether or not to engage and kill (Glas, 2015).

Working with its proprietary game engine, as an integral part of the design process for *The Last of Us*, developer Naughty Dog adopted its responsive AI technologies to heighten the impact of player choice to build a gameplay experience that carries with it the emotional weight of each enemy kill. Travis McIntosh, Programming Director at Naughty Dog, elaborates on this design calculus:

> When we started prototyping the human enemy AI, we began with this question: "How do we make the player believe that their enemies are real enough that they feel bad about killing them?" Answering that one question drove the entire design of the enemy AI.
>
> *(McIntosh, 2015, 419)*

Game development was in part a matter of solving an AI problem to build a range of NPC behaviors: intelligent and coordinated, or chaotic and aggressive. At the same time, the player's companion characters (NPCs that engage and interact with the player in distinctly different ways from other NPC classes) have their own behaviors, built to foster playable dependencies and deeper relationship building. Across the game's chapters, different types of AI characters fit both the needs of gameplay and the dramatic arc of the story.

With a suite of AI tools that were developed across the *Uncharted* series and have since evolved, Naughty Dog has defined a range of characters and behaviors and attached these to an underlying AI architecture, and it has maximized the profitability of its underlying technologies by iterating them across its two major video game franchises. AI is a functional property of a number of modular systems focused on distinct gameplay attributes that cut across multiple workflows—designer interface components that allow the central definition of roles and behaviors, aesthetic components that handle how these look and animate, and scripted decision-making components that determine movement, trajectory and performance (Russell, 2010). A set of default data is used to define a game's various AI systems and to create a series of archetypes that can be assigned to AI-enabled game objects; this modular data structure allows the features or skills of these archetypes to be easily modified with a custom scripting language to suit diverse object-oriented contexts (Russell, 2010), and it allows the engine's AI subsystem to be repeatedly developed and enhanced without destabilizing the overarching game engine architecture.

The Last of Us utilizes finite state machines, an AI technique I have discussed in the previous chapter. Finite state machines (FSMs) represent a long-established approach to crafting AI behaviors; they simulate sequential logic by outlining a fixed set of states. Only a single state can be active at any time, and each machine agent transitions from one state to another in response to a sequence of inputs that are attached to each state. FSMs provide a means to classify and scaffold individual intelligent behaviors as exclusive states (Thompson, 2020a)—for example, a character will attack a target or search an area until a system input (a game event such as the sound of an explosion or a visible change in the environment) tells the character to transition to another state. FSMs are not the only contemporary approach to game AI. Behavior trees can also be used to generate responsive in-game behaviors, but they use tasks instead of states, while minimax algorithms are commonly used in two player turn-based games with limited board states, such as chess. FSMs were put to notable use in *Half-Life* (released in 1998), Valve's first-person shooter, where they served to create acutely aware and situationally dependent enemy team behaviors. As I have already noted, FSMs have been used in some of the earliest arcade games to define simple yet distinct states. The ghosts in *Pac-Man* are FSMs that transition between three states, as they roam freely, evade or chase the player, and where the conditions for transitioning from one state to another are linked to the game's power pills. Valve introduced considerably more behavioral complexity into *Half-Life*, where NPCs can perform individually or collectively as they negotiate richly detailed and physically dense and varied environments to locate and overtake the player. While state machines are mathematical models of computation, they also serve as functional abstractions as states can have variable attachments; they can be assigned to any number of distinct behaviors, and those behaviors can be responsively connected to any number of external environmental factors (or inputs). FSMs provide a usefully simplistic logic to define actions and reactions, and they can significantly reduce the complexity of a game's underlying code. At the same time, the malleability of their assigned values allows them to be readily attached to most narrative containers; they simply require a bit of sequential thinking. The legacy of Valve's approach to AI is not only about how it functions to strengthen opponent intelligence but also its narrative association with combat, with goal-oriented tactics and with military strategy more broadly. In their situational analyses, AI systems read players as part of a game state and respond accordingly.

Valve's development of AI across its intellectual properties is notable; from the GoldSource engine released in 1998 to the Source engine which debuted in 2004, Valve has advanced the abilities of its enemy NPCs and by extension their dramatic adaptiveness. The artificial intelligence agent GLaDOS is the primary antagonist of Valve's *Portal* (released in 2007) and *Portal 2* (released in 2011), a series of narratively connected puzzle games that feature elaborate spatial physics; and *Left 4 Dead* (released in 2008), which was built from a new branch of the Source engine, features a situationally aware AI director that spawns enemies and items in relationship to the player's current status to dynamically shape the game's pacing and difficulty.

To increase their profit margins and streamline ongoing development work across any number of intellectual properties, Valve, Naughty Dog and other AAA developers are invested in iterating their game engines and the AI subsystems that are attached to those engines, advancing their engines while retaining fairly consistent models of computational intelligence. As Valve's work illustrates, studios make significant investments in their proprietary engines, which need to be repeatedly versioned or modified to keep pace with evolving industry standards. The code work needed to produce and maintain an engine is significant; to recoup those labor costs, engines need to have an extended lifespan. Most contemporary game engines are modular and built to evolve incrementally, and allow for different subsystems to be updated independently without breaking the engine's overall programming framework. Most versioning is attached to new game supports and features, some dictated by changes to system processors, operating systems and other industrial shifts, and others by internal choices such as the new forms of gameplay associated with specific game titles.

The most notable evolution of Naughty Dog's AI subsystems across two of its major franchises, *Uncharted* and *The Last of Us*, is how AI attributes and behaviors are embodied and performed, and have shifted the dynamics between character assets. More simply put, the narrative contexts of the studio's games have evolved, and though the AI subsystems have remained fairly constant in their code frameworks, their execution, their animations and their motivations have become increasingly dramatic, wrapped in plot and story and responsive character behaviors. From the first *Uncharted* (released in 2007) to *Uncharted 4: A Thief's End* (released in 2016), and between *The Last of Us* and its sequel, enemy AI has become increasingly environmentally driven, exhibiting appropriate grey zones or moments when an NPC may seem to be hovering between states and may give up on directional activities such as searching (Linneman, 2016). During these moments, NPCs lose awareness of the player and re-establish that awareness only when the player emerges from cover and re-enters their line of sight or creates an audible disturbance in the game world; if the player moves too quickly while attempting stealth, or breaks a bottle, brick, window or other object, nearby NPCs will take notice. While the AI of the *Uncharted* series allows NPCs to move between two states, between ambient behaviors while unaware of the player and combat behaviors once the player is discovered, the state machines of both installments of *The Last of Us* are far more complex, allowing NPCs to demonstrate ambient behaviors, investigate, combat and search for the player, and transition freely back and forth between these states (Gregory, 2014). Nonetheless, each of these game titles holds on to the fundamental reality principle of keeping characters in non-conflicted states and attaching state transitions to observable environmental forces or audible cues.

The AI of *The Last of Us* is built around skills (high-level character actions) and behaviors (the particular implementation of those actions), with the latter uniquely tied to each character type (Botta, 2015). Both Hunters and the Infected can move and attack, but they move and attack in different ways. The game's Hunters investigate environmental disturbances, hide behind cover and can adroitly navigate the game map to flank the player, while the Infected wander aimlessly around the map but, when alerted, will chase their opponents head-on. Each high-level character action is tied to a more specific and localized action as non-player characters respond to the changing state of the game world; as they move, react and interact with the player, NPCs do so in ways that are designed to look intelligent and that reflect each environmental context (Thompson, 2020a). Hunters may call out to each other as they patrol and search, and in doing so inadvertently communicate critical information on their intentions and their movements to the player.

Game behaviors are generalizable; while they reference specific concrete actions, these might be executed by every character. Despite their universal and recyclable nature, behaviors are performed and animated differently by each character type (Thompson, 2020a). That difference, that design intention, is built to be perceived as a means to inform the player's strategy against each NPC class. The modular structure of behaviors and skills allows these to be readily decoupled and independently iterated, as system developers pivot toward new design goals (Thompson, 2020a).

The Hunters of *The Last of Us* appear coordinated; they can detect player disturbances in the world, communicate their behavior to one another and consider their own personal safety. Their skills (panic, advance, melee, armed attack, hide, investigate and search) are built around combat and driven by player detection. Hunters rely on an underlying view cone for detecting the player in space; each NPC runs a raycast (a line-of-sight test) from their position to scan for the player's body and to identify occlusion points—any environmental features that would obstruct their field of view (McIntosh, 2015). A navigation map shows the NPC the fastest way to navigate the world and an exposure map shows the NPC what they can see from their current position to determine the safest path forward. The navigation and exposure maps are grids that sit on top of a navigation mesh, a governing data structure associated with each level design that allows NPC agents to calculate paths through the environment. Utilizing these data sets, the system generates a search map, which indicates which areas of the world are not yet visible and sends the NPCs there to explore. The NPC develops a rich understanding of the world, its cover points, and the location of the player, and by ranking a series of sequential raycast results, it selects the safest and most direct navigational paths (McIntosh, 2015); time-stamped data packets retain the location of the player and are regularly called out to continuously communicate and update the player's location and combat vector (the direction the player is currently facing) and reset the NPCs ideal position. This process makes tighter and more enclosed environments more hectic and dynamic, as these limited spaces produce more rapid communication cycles and make it more difficult to lose the enemy; open outdoor environments or larger combat spaces are less frenetic, as they provide more opportunities for players to evade NPCs.

There are unique classes of Infected NPCs whose skills and sensory systems differ from one another, but sound generated by player movements or in-world objects (such as generators and vehicles) is the primary trigger for all of them. The radius of the sound event can be multiplied by a tunable value for each character type, to scale with the speed of the player's movement and the specific sensitivities of each Infected class. There are also a number of low-level sound events in the game world (such as the natural breathing of the game's playable characters) that are designed to help the Infected find the player if they are hiding in close proximity. To counteract these low-level sounds, players can throw bricks and bottles to create audio distractions, or craft Molotov cocktails and smoke bombs to kill or deter the Infected. By default, the Infected will wander, visiting a series of interaction points on the map until they are distracted by a noise or nearby combat; this will trigger their more directed search activity, which is focused on locating and moving toward a disturbance and canvassing the surrounding area until they settle back into wandering (Botta, 2015). As they begin to understand the motivations and movements of each NPC class, through observation and repetition, players are able to develop better survival strategies. Players are rewarded for moving slowly and deliberatively through many of the game's environments, and keeping their noise level to a minimum. To counterbalance these strategies, with an AI system capable of nuanced contextual decision-making, the NPC is able to understand and analyze the field of play and adapt to the player's actions (McIntosh, 2015).

Behaviors and skills may be properties of the game engine and rooted in its codebase, but they gain meaning through their physical embodiment in distinct character classes. This is not simply a matter of information visualization, but of data physicalization, as data define the NPC as a physical body that can move through the negative space of the game world; character design and level design are interdependent data constructs filled with interrelated potential (Offenhuber, 2020). NPCs are, in part, defined by their environments and the player's experience of the environment, as safe or threatening, is shaped by the presence or absence of NPCs. Bodies and spaces are designed interpretations of a collective and unified data set, although that data set is itself multilayered, connected to the processes of engine-based mechanics as well as the assets of the game world, which represent and approximate material processes.

While the developers at Naughty Dog have imbued the human NPCs of *The Last of Us* with a high degree of emotional intelligence, often expressed through their verbal communication, generic adversaries remain first and foremost a design problem, an obstacle for the player to evade or overcome, and an integral part of a larger environmental puzzle conceived during level design. By opening up the game world, allowing for open-ended problem-solving and building more nuanced AI systems, game developers are able to increase player agency and introduce ethical analysis and decision-making (Glas, 2015). Nonetheless, game AI is a system, a component of an engine and a foundation for game mechanics; and as a system, for large-scale developers such as Naughty Dog, it is a vital part of a serialized intellectual property and has by necessity coalesced into a repeatable yet expandable design pattern; Naughty Dog's AI system is a reliable problem-solving method and a reliable method for staging in-game problems. While an AI system is a structural tool used to guide the development process, it is also a design language with a well-tested logic that determines the playable potential of each game. AI systems define the behaviors of NPC agents and they dictate to a large degree how players can affect those agents. Throughout *The Last of Us*, the AI system provides a cohesive knowledge structure, grounded in the anxious architecture of a pandemic world, that guides design, analysis and play (Björk and Holopainen, 2006).

Pandemic Culture

The Centers for Disease Control drew a humorous bridge between fictional and nonfictional pandemics in May 2011, when Dr. Ali S. Khan, the Director of the Office of Public Health Preparedness and Response, authored a blog post titled "Preparedness 101: Zombie Apocalypse" and discussed within it the ways the CDC would respond to a zombie attack. The lasting relevance of Khan's post speaks to the plasticity of the zombie metaphor. Zombies are a rather fluid signifier, attached to a series of cultural anxieties about the loss of self and control (to others, to technology, to capitalism, to communism, to globalism and to generalized terrorism and wide-scale destruction) and perceived threats to the social order. Neil Gerlach and Sheryl Hamilton (2014) note, "more immediately and fundamentally, viral zombies are about apocalyptic levels of disease and contagion." While pandemics tend to reify existing racial and socioeconomic inequities (notably, the uneven access to quality healthcare), viral vectors inevitably attach every person to the collective; moreover pandemics, unlike localized outbreaks, leave a permanent imprint on society; they point out the vulnerabilities in our social infrastructures and the penetrability of our personal borders and national boundaries, and they lead to the adoption and erection of new protections, new barriers and new practices. In this landscape, zombies operate as vessels that make viruses visible (Gerlach and Hamilton, 2014), and they provide a productive counternarrative about the loss of self-management and, in a time of COVID-19, the risks associated with not wearing a mask, washing our hands or staying at a safe distance from others.

As with the primacy of NPC AI in survival video games, pandemic culture fosters parallel dependencies toward pattern recognition, to seeing contagion as a curve and to mapping social distance across a number of both localized and generalized demographic, geographic and physiological vectors. Our actions during the Coronavirus pandemic are regularly influenced by the physicality of data. As we travel (or choose not to travel), we are guided by our perceptions of particulate material vectors. We can draft an analogous assessment of game logic and the everyday logic of the algorithm if we come to understand how agency operates (through data) in times of public crises and how that operational feat does not occur in isolation but instead follows other patterns of informatic design. The Coronavirus pandemic has forced us to see public space through a new lens, to see indoor and outdoor environments as a series of traffic patterns and safe distances, some marked by CDC-recommended signage; as we navigate these spaces with a heightened sensitivity of what is possible, we become acutely aware of those around us. We see them as potential intruders in our six-foot radius. Managing the pandemic means understanding and finding comfort in epidemiological data: the mechanics of viral spread, the data-centric determinants of social distancing and protective wear, the viral half-life of surfaces, and the long-term arc of containment through testing, contact tracing and immunization. Managing the pandemic means translating the science into a data-driven projection or a non-threatening symbolic agent (six-feet measured in a cow's length) that allows us to act with purpose. The comfort of data is making the unknown knowable, making the invisible visible, defining acceptable levels of risk and establishing agency. Data are the necessary ground for evidentiary logic and for the statistical modeling used to set policy and take action. While we might suggest that society's interaction with data-driven decision-making during a pandemic follows longstanding interactional and organizational practices that have

FIGURE 3.1   From the Centers for Disease Control and Prevention poster, "Help Protect Yourself and Others in Public Settings."

Source: Copyright 2020. Centers for Disease Control and Prevention

been similarly driven by data and facilitated through artificial intelligence (such as traffic planning, urban development and predictive policing), pandemics create new problems, new logic models, new urgencies and new conditions that require the coordination of otherwise discrete agencies and industries and the integration of otherwise discrete data sets and resources, often at grander scales.

Artificial intelligence has been leveraged for detection and disease management throughout the COVID-19 pandemic and has performed similar work during a number of earlier public health crises. Outbreak detection, control and operations rely on a number of automated surveillance strategies, enhanced by machine learning, to extract meaning from localized community data. AI has been used to identify disease clusters, monitor case numbers, predict future outbreaks, outline mortality risks, allocate resources, advance vaccine research and the development of therapeutics, and engage in ongoing record reviews to recognize patterns and understand disease trends (Arora et al., 2020); and AI has spawned a new sector of public health technologies. AI has been integrated into mobile health applications to improve diagnostics and advance contact tracing, where it has furthered global efforts to combat COVID-19, but the attachment of AI to sensitive and locative consumer data has presented new challenges to widely held commitments to privacy, destabilizing recognized civil liberties (Leslie, 2020). China led a first wave of digital health surveillance by using mobile phone data to regulate the movement of its citizens. The Alipay Health Code system, a government project first introduced in the eastern city of Hangzhou, delimits public sector mobility and identifies those who need to quarantine (Mozur et al., 2020); by integrating health, travel status and location data, the application generates an individualized QR health code—stratified into green, amber and red levels—that determines whether or not the holder can travel freely throughout the region. In the United States, Google and Apple joined forces to create a parallel Exposure Notifications framework that acts as a form of digital contact tracing based on event detection; the opt-in system setting (available only where it is supported by regional public health authorities) uses Bluetooth to notify users of potential exposures to other users who have presented an active COVID-19 diagnosis. The framework calculates an Exposure Risk Value determined from a number of interrelated data structures: transmission risk (the status of the infection in the affected user), cumulative duration of exposure, days since exposure and attenuation as an approximation of physical distance during exposure (Apple Developer, 2020). These values determine an overall risk score for each exposure incident. The framework translates infected bodies into networked viral vectors, calculates risk from symptoms, space and time, and bifurcates individuals into two roles: affected and potentially exposed users. The distances between people have meaning and, as I have suggested, bodies and environments can be treated as functional data sets. COVID-19 contact tracing applications that read contact events as equations of time and proximity depend on the real-world performance of Bluetooth, where signal propagation can be complicated by various physical materials. In their measurement study of signal attenuation on a commuter bus, Douglas Leith and Stephen Farrell (2021) note,

> the propagation of radio signals in practice is often complex, especially in indoor environments where walls, floors, ceiling, furniture etc. can absorb/reflect radio waves and so change the received signal strength. A person's body also absorbs Bluetooth LE radio signals so that the received signal strength can be substantially reduced if their body lies on the path between the transmitter and receiver.
>
> *(2)*

As a model for collecting field-based exposure data and mapping viral vectors in a lived context, bodies and objects are treated as materials and are unilaterally reduced to their properties of attenuation.

It is important to clarify that the AI used to manage a public health crisis is of a different order than game-based AI, at least as the latter is commonly focused on the appearance of intelligence and a localized form of decision-making within a discrete set of organized parameters—parameters whose variables have been largely pre-determined by the game, its engine, programming, design and world rules. While the statistical models developed by public health agencies are also parameterized, they are not bound to the limited rules of industrial design and development. Moreover, the patterns of game-based AI are tied to the behaviors of technologically determinate subjects and objects, rather than the relatively unbounded categories of human subjects. As well, game AI is a readable and decipherable system; it is knowingly attached to an underlying logic. Players understand game AI at work and do not confuse it with true intelligence, although they may attach greater intelligence to an AI agent than is warranted and established in its code. Of course, the orders of AI that are associated with game- and non-game systems are often intermingled within a particular application space, and the distinctions between them are slowly but surely being erased, as game developers look to establish more gratifying forms of player agency and as the tenets and technologies of game development come to undergird a much broader range of media architectures; the cross-industrial applications of artificially intelligent engine-based interaction have been activated across an ever-expanding number of industries that include healthcare, transportation, manufacturing and urban development. Moreover, game and non-game AI systems interweave bodies and environments; they account for biological, spatial, temporal and material structures, and understand these structures as data points, as inputs, to calculate an output and an appropriate course of action.

Crises reveal precarity and uncertainty; they speak to evolving psychosocial and material relations, and they lead to commercially and politically bound narratives that serve to reconnect or reconstitute society in relation to the present and the future, as a mediated form of therapeutic intervention. Data, tethered to artificial intelligence and to algorithmic projection, play a significant role in crisis management; data can be harnessed and filtered to provide situational awareness, visualize risk, generate tactical insight, optimize the distribution of physical resources, and meet the universal linguistic demands of crisis informatics systems. Crises are not necessarily contained by such machinations, but our attachments to technology fuel an economics of production that provides a ground, a logic and a way of organizing and making sense of the world again. This process grabs hold of social space and aligns its practices with its representations (Lefebvre, 1991, 38). Pandemics tend to upend historical patterns, and in doing so, they destabilize longstanding algorithmic practices, but they also produce new methods of being in the world, and they produce new algorithms that can resolve previously unformulated conditions. As an example, the COVID-19 pandemic crippled the reliability of airline industry algorithms. Before the pandemic, airlines used algorithms to predict ticket demand based on historical data trends; yet these patterns were undermined by the pandemic as travel slowed, restrictions were set in place, and widespread cancellations undercut the reliability of even live data (Sindreu, 2021). In response, the airlines developed new forecasting models to rebalance ticket costs and stabilize their revenues. The COVID-19 pandemic introduced uncertainty into past patterns of mobility, but that uncertainty eventually yielded to new risk reduction methods and new models of individual and communal behavior.

Dark Play and Dark Tourism: Pandemic Behaviors in *The Last of Us*

In *The Last of Us*, players guide Joel and Ellie, the video game's two central protagonists, through a decaying post-pandemic world overrun by a brain-altering fungal virus that has swept the globe and transformed humans into mindless infected creatures, while survivors have divided into warring factions with their own territories. The survival game is driven by human stories, framed by cinematic and environmental spectacle and dynamized by an artificial intelligence system that weaves these elements together. *The Last of Us* is a third-person action-adventure game with survival horror elements and a tactical focus on cover shooting and stealth across varied natural and built terrain. As players make their way across post-apocalyptic America, traversing urban, suburban and rural spaces, they come into contact with two types of opposing forces: Hunters, the humans who patrol and control regions of territory around the country, and the Infected, the mindless violent creatures that are all that remains of those consumed by the contagion. In the first game, players cross multiple state lines—Texas, Massachusetts, Pennsylvania, Wyoming, Colorado and Utah—and wind their way through major metropolitan areas—Austin, Boston, Pittsburgh, Jackson, Boulder and Salt Lake City. The second game has a similar reach; while the story opens in Jackson, Wyoming, the major events unfold in Seattle over a three-day arc (repeated from the perspective of two dramatically entwined characters) and climax in Santa Barbara. These cities are dissected and translated into functional level designs, engineered as informatic environments that propel or hinder movement, set conditions on progression, establish rules for traversal (by attaching the appropriate software code, compiled as a traversal action pack, to environmental features and assets) and create objectives; game art and design converge to provide environmental cues, such as lighting and leading lines, that guide the player through the game world. Re-inscribed by the shorthand of efficient, functional game design, these cities are experienced first, from a distance, cinematically, and then up close as a series of environmental puzzles to be solved. The post-apocalyptic American city is scripted, programmed and designed as a narrative space governed by informatic exchange, and the pathway forward, the suggested ways of progressing through *The Last of Us* are purposefully dotted with points to pause and on occasion converse with other characters; while image-making is not a formal plot device, the level designers have thought about vantage points, about the purposeful framing of the game world through a series of well-articulated in-game camera positions that highlight how nature is slowly reclaiming the built environment. In a particularly dramatic sequence in the first game, Joel and Ellie pause to watch a herd of giraffes roaming through the deserted ruins of a fallen city. Within these purposefully engineered resting points (which recall Kodak picture spots), players are able to reframe the scene by guiding the game camera through their avatar. Active looking is encouraged throughout the game, and players are trained to do so, but here, where character movement gives way to camera movement and the right analog stick dominates the left, game art naturalizes the game's code work. When they are on the move, players are tasked with investigating their surroundings to find hidden supplies and evade enemies and remain much more attuned to the game's mechanics; players are rewarded for exhaustive looking, for discovering optional conversations that add psychological depth to the game's characters, and for finding hidden collectibles, health items, and other resources

FIGURE 3.2   A map of downtown Seattle provides direction for Ellie and her non-player companion character Jesse, as well as a series of waymarks for the player in *The Last of Us Part II.*

Source: Copyright 2020. Naughty Dog

that can advance the player's skills and weapons. These acquisitions are realized as expressions of health, skill and firepower in the avatar's executable code that change the dynamics of gameplay. In *The Last of Us*, looking, reframing and only then moving allow players to visualize and realize different outcomes (Zylinska, 2020); players are rewarded for studying the environment, identifying possible obstacles and discovering new affordances before they continue onward.

Video game space is coded, manufactured, populated and realized through a number of game rules and mechanics to control the flow of information and ideas, which form the calculus for player navigation and movement; these rules and mechanics form the foundation of action, and movement is continuously monitored as part of an informatic loop that defines the state of play. But fully realized code must also communicate the possibilities of a space and relay the latent potential of its design. While the in-game mechanic forms the primary horizon of the player's engagement, it too depends on an environmental context. A form of emotional projection binds together a game's central protagonists, but the same type of projection does not commonly extend to those bodies that, although similarly inhabited by AI, are defined by their appearances and actions as enemies. These bodies are read differently. As they progress, players are continuously reading the environment, negotiating its inhabitants and executing a number of context-dependent commands. The city is a site of organized possibility that is always complicated by algorithmic unpredictability and expression. Studio-produced gameplay modifiers (unlocked alongside the Game Plus mode of *The*

Last of Us Part II) such as slow motion, infinite ammo, infinite crafting, one shot and touch of death tamp down the level of unpredictability and alter the balance of power by increasing the player's level of certainty and control. These gameplay adjustments and aesthetic modifications change the game's degree of difficulty and enhance its replay value. Outside of these modifications, to progress, players must consider the game world holistically; despite the abundance of its code work, players must understand all of its assets and mechanics and recognize its patterns. These patterns become more readily decipherable with each playthrough.

The cities of *The Last of Us* are part of a more widespread speculative form of geographic projection that asks us, from a current state of crisis, to forecast a collective future. They function as pandemic porn; they express sweeping societal collapse through the spectacle of diseased bodies and deteriorated structures, and they open up an uncanny valley by visualizing the incongruity between the present and the future. We experience these cities as a third-person projection that can somehow traverse an otherwise uninhabitable universe. The cities of *The Last of Us* form a subset of the pornography of disaster and the spectacle of suffering and catastrophe, made real in the voyeuristic experiences of dark tourism, a real-world practice of visiting places of impossibly imagined disaster or of protracted abandonment and decay, where vibrant structures have fallen into brutalist ruins. The game world is a space consciously constructed as an "affective socio-spatial encounter" (Martini and Buda, 2020, 679) compounded by the anxiety of reaching checkpoints to save the game state and mark progress while combatting a series of distressed bodies that function as powerful distillations of trauma; this segmented space, bifurcated by events, gates that stop progress and valves that prevent backtracking, is hyper-analytically detached from its whole. Dark places are often unruly accretions of nature and manufacture, inhabited by our impressions of the distant and near past and activated by our efforts to understand what happened; these mental efforts create inflection points, produce moments of heightened awareness along an imagined timeline. Throughout *The Last of Us*, these are given form as punctuated encounters with monstrous others and animated displays of bodily harm and decay. Dark places, as sites of uncertain trauma, open up a distinct process of reading, reflection and integration; they must be performed and ultimately over-written with new narratives that create a safer space for the traveler. But dark places can be developed and engineered, manipulated to form a coherent and artificial narrative (Martini and Buda, 2020, 686). The overgrown Seattle of *The Last of Us Part II* is both Seattle and "not-Seattle." The Seattle Waterfront Aquarium is a location featured in the game, and a key site of narrative tension, but as with most set pieces drawn from the city's actual architecture (including the downtown courthouse and a catalogue of other physical structures that act as containers for play), its decay-enhanced volume gives way to an even more deeply reimagined interior life, architected as a series of interlocked play-centric spaces with underlying markups that indicate connectivity. The basement of the Lakehill Seattle Hospital, the arena for Abby's battle against the Rat King (a monster composed of multiple Infected classes that have been fused together), is designed as a series of cramped interlocked spaces. The player's strategy is shaped by this design which encourages limited frontal assaults before retreating into the darkly lit maze of abandoned hospital rooms. Players encounter this boss after dropping down from the floor above and are presented with no means of backtracking through the level.

FIGURE 3.3 Blockmesh is used to establish level flow as part of game design and development in *The Last of Us Part II*.

Source: Copyright 2020. Naughty Dog

FIGURE 3.4 As Ellie, the player can utilize stealth while reading the movements of enemy non-player characters in *The Last of Us Part II*.

Source: Copyright 2020. Naughty Dog

The functional approach to the re-imagined game-enveloped city draws from the real-world architectural principles of building information management, a data-centric model-based process that makes buildings and environments actionable (scalable, traversable), designed with a consistent and suggestive language (of shapes, colors, landmarks, lighting and leading lines) that communicates affordance and guides the player. Playable indoor environments must suggest and support (or deny) a range of specific tactics that allow the player to navigate the monstrous threats that make unforgiving interior spaces so hazardous.

Pandemics can rewrite most enclosed spaces and assign them a level of risk based on the likelihood of infection. Airports, city buses, classrooms, gymnasiums, markets, office buildings and public parks become spaces that must be approached with caution; outbreaks transform them into statistical models (grounded in the probabilities of who and how many people may be there) and sites of anxiety. The world comes to embody the disease characteristics of a pandemic, activates mental projections of contagion, imbues objects and individuals with the adverse power of endangerment (Aaltola, 2012). The pandemic settles into the surfaces and substrates of the object world. In this manner, game art, level design and mechanics are all calibrated expressions of pandemic anxiety.

Navigating With Intelligence

AI is not only used to map, define and understand the possibilities of a space; it is also used as a navigational tool and as a mechanism for action. AI is connected to the coded rules of player engagement; while the movement of the player in space, the choice of how to navigate and interact with the environment and with NPCs may happen with a relative degree of openness, it always hinges on a bounded body, itself a code object whose rules are not governed by the player, but by a series of coded attributes, of boundaries, of transformations, of physics that are algorithmically transformable. As a guiding philosophy brought into practice, Naughty Dog developers have pursued a balance between authored (scripted) content and systemic behaviors; while combat in the *Uncharted* series has historically been tightly authored by the design team, *Uncharted 4: A Thief's End* reflects a significant turn toward a more systems-driven approach to artificial intelligence to support expanded gameplay in large complex environments (Gallant, 2017).

In *The Last of Us* and its sequel, the AI systems that control NPCs run parallel to companion AI systems that bring Ellie, Joel, Dina, Jesse, Tommy, Lev, Yara, Manny, Mel and other principal characters to life. While Joel and Ellie function as the primary playable characters across the two campaigns (Abby joins as a critical playable protagonist in the second installment of the franchise), they are also cast as AI-controlled companions defined by their utility to identify and attack enemies and protect the player. Companion characters maintain a reasonable proximity to the player and find points in the world that make sense for them to do so, as the game builds a follow region that guides companion AI. When the player goes into cover, the AI system draws from a series of raycasts generated from the player's position to locate logical environmental cover points, prioritized by their distance to the player and to nearby threats. Drawing from this script, companion characters are imbued with a sense of agency as they react to the player and to events in the world.

Playable characters, companion NPCs and antagonistic (and supporting) NPCs are part of an ecosystem and their movements and expressions adhere to a dominant pattern language, and all must be able to reason well while navigating complex three-dimensional environments. The *Uncharted* series makes significant use of vertical combat spaces, while *The Last of Us Part II* utilizes a wide range of axial states—crawling and prone—that add to the game's geometric complexity. Characters are woven into the fabric of cities, as they are abstracted and parsed into a choreographed level design, and their scripting must support the ability to read the terrain and perform the appropriate traversal actions and animations. This is the matter of responsive worldbuilding, of connecting higher- and lower-level orders of design, and it suggests a careful attention to balanced and situated practices that establish a recognizable navigational outline for both the player and the NPC while allowing the city and other spaces and landscapes to be dynamic and discoverable. Pandemic narratives activate and exploit these entanglements; they are scripted to call out the interrelated embodiments of players and NPCs. These respective object classes remain acutely aware of each other and function together in the game state to translate the spectacle of the pandemic into a series of intimate contact points; the tactical engagements between players and NPCs

FIGURE 3.5　Blockmesh is also used to delineate and evaluate pathways through the physical environment as part of game design and development in *The Last of Us Part II*.

Source: Copyright 2020. Naughty Dog

mirror the very real fact that contagion and spread depend on the dynamic inter-action of objects in the world (Aaltola, 2012). Players survive by building mental projections of each level, considering the necessary resources, and navigating a series of strategic trajectories through the game world. Players must come to understand the game's relative authority, and understand how they may oper-ate within the scripted flow of the pandemic spectacle (Aaltola, 2012). Pattern recognition is one form of pleasure in playable media, and a particular design strategy that often produces a balance between pre-determined algorithmic logic that can be readily perceived and exploited by players, and fuzzy logic that is more nuanced, adaptive and challenging, and impedes players from gaining direct insight into a game's abstracted processes.

The character dialogue systems for *The Last of Us* are built with contextual awareness and are an audible trace of the game's underlying AI subsystem. They provide signposts for the player to understand the game's underlying intelligence frameworks, but that awareness is designed to create heightened realism as it highlights the contextually driven motivations of each NPC. The AI systems of *The Last of Us* and its sequel allow NPCs to converse with each other and allow companion characters to connect with the player, and they provide an overt communication loop between all sets of characters to pro-duce actionable information for the player. In his overview of the dialogue requirements of *The Last of Us*, Jason Gregory (2014), a Lead Programmer for Naughty Dog, notes the interdependencies between the game's dialogue system and its AI system, both as a feat of programming and as an attribute of gameplay; the dialogue system is responsible for openly telegraphing AI state changes, letting the player know whether NPCs are on alert, creating dynamic assistive conversations during combat, reinforcing the helpfulness of compan-ion characters, providing location cues for enemies, showing intelligence and coordination among NPCs, reacting to player actions, establishing when NPCs begin to search, letting the player know when stealth is broken, giving charac-ters personality and assisting in telling the story (Gregory, 2014). The dialogue system also allows for interruption and can dynamically shift a conversation to respond to changes in the game state as a result of the player's actions. The dialogue system is part of an event-driven scripting system that can translate a gameplay event into a physical line of dialogue (Gregory, 2014). Character animation and dialogue systems are interrelated signatures of AI programming, although the former is commonly procedurally generated, while the latter draws from a store of pre-recorded audio files. While intelligence may signal how NPCs act and allow players to read those contextual motivations, the visual and aural traces of what AI is doing are equally relevant to sustaining a game's narrative drive and plausibility. Programming ties these strands together and integrates them into the appropriate locations in a game's AI code; and the labors of audio designers who can write simple rules to select between alternate lines of dialogue during runtime are integrated into the game's overall code work, reduced into basic data types (as part of a fact dictionary or information store) to call out functions from its engine while also accounting for context (Gregory, 2014).

When AI is deployed in the service of realism, it needs to remain grounded in the environment; to this end, code also concretizes, normalizes and conceals its procedural imperatives, and only reveals its embodied traces. While not an open-world experience, the settings of *The Last of Us* and its sequel provide the player with options on how to proceed, options that are contoured by the environment and the presence of intelligent AI systems that allow NPCs to analyze playable space and the player's movement through that space to understand the current state of the environment (Botta, 2015). The dystopian plot framework of *The Last of Us* juxtaposes confining city spaces (that seem ground zero for the post-apocalyptic crisis) with the open spaces and potentialities of the natural world, a dichotomy outlined by Gerald Farca and Charlotte Ladevèze (2016). But both spaces are equally threatening, requiring the same skillful navigation and mastery of the terrain and the AI systems; these spaces are engineered for parallel deployments of stealth and combat, and their varied environmental features (tall grasses, built structures, scalable objects, breakable bottles and bricks) act as cues for progression. Moreover, these cues are tied to the graphics features of the in-house game engine that offer fidelitous physically based material rendering and enhanced environmental lighting (a blend of baked and runtime lights) to highlight the form and function of terrain.

The survival mode of *Uncharted 4: A Thief's End*, the last installment in a serialized intellectual property also developed by Naughty Dog, extends the game's AI systems into a cooperative combat mode (the game also includes a competitive multiplayer mode), which over time has been studied, learned and shared among players to develop time-saving strategies. Mapping AI requires understanding how NPCs spawn, move and navigate level maps in a dynamic field of play; advanced players have identified these as a number of pattern-based waves that are to varying degrees randomized depending on team-based player interactions. These lessons are often codified in published walkthroughs and playthrough videos (on *IGN*, *Eurogamer*, *GameFAQs*, independent YouTube channels and other online sites) that lay out strategies and identify significant level map features such as choke points and points of cover. Advanced players are able to read both character and environment and balance the speedy reading of multiple spatial representations—from the three-dimensional third-person playable perspective to the two-dimensional icon-based mini-map (part of the heads-up display)—that are abstractions of an underlying codebase that includes the AI framework. Survival mode also highlights the various uses of non-player characters, as always-available human-like opponents that execute human-like play styles, and as always-available human-like teammates that execute rational cooperative behaviors. Players can tackle survival mode with a companion AI or with a team of players. These experiences are designed to be interchangeable; players might start a match with an NPC and be joined mid-match by another player, forcing the NPC to drop out.

Survival Lessons

There are underlying operating behaviors that govern and produce co-dependencies between playable and non-playable characters. The data needed for AI to function is poured into a simple structure that commonly foregrounds a game object's abilities: a list of behaviors, skills and character classes that draws from the mutual work of programmers, designers and animators. And these potentials are designed to be game-dependent, experienced by individual players but responsive to player characters. Follow systems, for example, are fanned out from the player character's position; at the same time, for companion NPCs to properly follow or lead the way for the player, they must incorporate a number of clearance tests that are environmentally dependent. Level and character design may be bifurcated workflows in the game development pipeline, but they are executed through a common data set.

Level design is an algorithmic approximation that requires managing horizontal and vertical space, providing rational geometries that influence play and offer environmental reference points—landmarks that stand apart from the general terrain, destination points that are distinct from other general cover points scattered

FIGURE 3.6   The buddy follow system of *Uncharted 4: A Thief's End* lays out potential positions for non-player companion characters in relationship to the player character. Each position must satisfy a number of clearance tests.

Source: http://allenchou.net/2016/05/a-brain-dump-of-what-i-worked-on-for-uncharted-4/

Copyright 2016. Naughty Dog

throughout the level, and transverse spaces that are distinct and functionally different from open geographies. Cooperative survival maps operate differently from their story mode counterparts; repeatable, re-playable spaces are less dependent on narrative and the visual cues (leading lines, lighting) otherwise required to guide the player through an unknown space that functions in part to advance the story. Exploration gives way to rote memorization, to understanding in rather precise ways the rules that govern a space and the strategies fundamental to survival. Through the distinct processes and technologies of game-based worldbuilding, programmers and designers have forged a paradigm shift in spatial mapping, leveraging software to produce new ways of organizing the world and the people within it, and to regulate player agency. The embedded lessons of predictability are tied to understanding algorithms and can help us flatten the curve and regulate the chaos of what may seem to be random in-game events. These lessons are conveyed through a linguistic structure that inexperienced players can learn over time.

The same holds true in cooperative multiplayer survival modes, as players can make swifter progress if they read and assist each other and work collaboratively against a field of common enemies. Team-based play can prove more challenging for players who do not possess the field literacy necessary to anticipate the actions of other bodies; it is common for players to drop out of an *Uncharted 4* survival mode match if their teammates fail to understand the level map's patterns and rules or fail to assist (and revive) each other.

The alert system of *Uncharted 4* is yet another novel communicative framework, a familiar diamond-shaped icon that hovers above each non-player character when a certain level of awareness is prompted by the player's actions. The icon has a variable alert level that slowly fills in as white and then changes color from yellow to orange as the threat level increases. The alert is triggered by sight—seeing the player character or seeing bodies of other NPCs that have been disposed of by the player character. The alert also triggers other spoken cues, as the NPCs begin to communicate with each other, alerting the player of their heightened state and their intent (to seek out and do harm). These verbal cues are not actually used to communicate with other AI characters (instead, this is accomplished by passing messages through the code), but rather to explain their actions to the player (Dill, 2014, 5). *The Last of Us* amplifies its communication systems as part of the development of dynamic stealth and as a corrective to the player's advantageous listen mode that allows Joel and Ellie to see enemies behind walls and objects and around corners. When enemy characters are searching for Joel and Ellie, they will know about each other, look for one another, and express panic (often they will call out to each other) when under duress. Sound and vision work in tandem as part of the game's threat-level index, and the orchestration rises and falls to match the range of dramatic tension, and fades to suggest that the coast is clear.

It would be shortsighted to discuss *The Last of Us* and *The Last of Us Part II* as data-based engine-driven properties rife with (visual and acoustic) algorithmic loops that require players to recognize patterns and navigate space. Yet playable characters and NPCs are both algorithmically driven translations of bodily dislocation, monitored by the player through visual and aural signposts and through the sensory-motor feedback of the game controller, that produce a complex chain of coded articulations, of movement and response across a field of play. The chain of player and companion movements includes an intimate degree of hysteresis, a lag in response between their respective behaviors; companions only switch states when the player has switched states (e.g., climbing and not-climbing states, in-cover and out-of-cover states) and moved far enough along while maintaining the current state (Chou, 2016). The rules of interaction, leading, following, distancing and patterning movement that govern companion systems are formed by code and placed by designers, and can be turned on and off in the design stage using a script. But narrative bodies are more than code objects; as with other forms of playable media, game AI gains an affective dimension when it moves out of design and development and into the field of play.

The Last of Us Part II must be played through male and female bodies—through Joel (in prologue), and Ellie and Abby (in parallel chapters)—but not through queerly open and mutating bodies (although the game includes openly queer and transgender characters—Ellie and Lev). The game's focus remains on progression, on forward movement across a factional society. The story unfolds as a number of boundary lessons. Consistently, players have to consider how to safely navigate open and closed spaces without being detected and attacked by radicalized (physiologically, morally) others: the Washington Liberation Front (and their guard dogs), the Seraphites, the Rattlers and various mutated classes. And players have to learn how to leverage their companions within each level map. There is a certain ludonarrative dissonance here that is common to many contemporary survival games—a tension between complex storytelling and the requirements of advancing gameplay that is bound to the game's mechanics. Ellie is cast as a playable and non-playable body; she is both protagonist and antagonist, and occupies that latter role in a brutal confrontation with Abby that closes out the Seattle chapters. But this dissonance smartly exposes the player's desire to occupy Ellie, even as her body is brutalized; this narratively motivated character exchange also allows the player to have situated awareness of more than one character throughout the game, and experience that awareness as it is abandoned, adopted and enacted by an AI agent. With its interwoven chapters told from two opposing functional perspectives, the game complicates the relationship between systems and story, as it tasks players with caring for and playing through both Abby and Ellie, and with considering the welfare of more than one body. Even during the game's final conflict, as an exhausted Ellie and Abby engage in hand-to-hand combat on a Santa Barbara beach, players are charged with acting and reacting to two oppositional yet complementary characters; having settled back into Ellie's playable body, players are tasked with freeing Abby from captivity (Ellie finds her tied to a post), following her down a beachside trail and, in a last antagonistic impulse, fighting her. As the shoreline battle unfolds, players have to execute a series of quick time events, knife strikes, dodges and weaponless melees before pressing the final "Strike" command. While presented with a confrontation that cannot be avoided, in a battle that can only reach one successful conclusion (the failure to execute the prescribed series of button presses and properly timed dodges and counterattacks leads to a narrative loop), players must consider the welfare of a previously occupied body while negotiating the limits of their own play state. Ellie's state of duress, she is bleeding and bruised, is conveyed through purposefully attenuated character movements, both as she trails Abby to the water's edge and as she melees with depleted force. Players must care for, rescue and battle Abby before they can finally let her go free.

Playable Media and Pandemic Performance

Throughout our current pandemic, data have been visualized to warn, inform and educate the public (Cooley, 2021), to activate the collective, and to focus our behaviors on combatting a common threat. The latent potential of data visualization is understanding how to act, and how to recontour behavior. In game design and development, data literally shape the visible field, and the mechanisms of engine-based production and artificial intelligence systems introduce a bias toward particular types of visualization that follow the imperatives of logical connectivity. Game development is governed by these ideals; it is regulated by a design sensibility (and a management and labor structure) that looks to solve a finite roster of repeatable problems and coalesces as a regime of practice, a limited number of organized patterns that structure our engagement. This is an exercise of governance by design, mandated and secured by the practical and experienced need for a rational worldview (Flyverbom et al., 2017) that nevertheless aspires to produce meaning beyond a game's mechanical interactions. Through the analogic intelligence of their NPCs and their reflexive construction of reality, games that advance machine intelligence attempt to more openly negotiate between the virtual and the physical world, the player and the video game, and allow these two spheres to influence each other (Keogh, 2018). To survive through *The Last of Us Part II*, players must occupy multiple intersecting bodies while monitoring the bodies of others. But beyond identification, beyond narrative, as an arena of play, the game is a lesson in mobility within a post-pandemic framework as well as an opportunity to understand the limits of overly determined playable media systems. Where the game falls short of the complexities of the pandemic world outside our windows is in its failure to invite subversive play, to allow play outside of its governing goal-oriented structures and to allow for open material negotiations that might violate the game's rules (Wilson, 2011)—the messiness that so often interferes with any unified pandemic response. Perhaps the game is simply calling out the limits of mediated embodiment that are written by design from the organizing architectures (both software and hardware) of playable media, where pathfinding (determining how to proceed within a set of environmental rules, such as choosing stealth or combat) is one of very few forms of expressive openness in AAA games. To experience the pandemic of *The Last of Us* is to bring to that experience what Brendan Keogh (2018) suggests is a "preconscious perception" of its world, already shaped by a number of objects, instruments and habits—among these, our collected experience with the technical realities and formalized abstractions of gameplay. This leaves us to consider the habits that shape our responses, as fully realized bodies with boundaries, to the pandemics that actually threaten us.

Local guidance for navigating public spaces during the Coronavirus pandemic urges us to plan ahead, be prepared and to keep moving. The longevity of the pandemic has raised certain questions about the return of urban life, as we once knew it. Public spaces have acquired new meaning; they have been populated with new signage, stamped with new directional flows and measured with new ratings systems. Practicality has supplanted sociability and meaningful inefficiency. The organizing principles of game culture have, for the moment, taken hold of civic life though they have not displaced its organic nature. There is a certain persistence of code, to the degree that it acts a bridge between algorithms and material outcomes. Code is the mechanism for regulating information and is designed to generate results from a known series of interactions; it both governs and makes us feel safe. It delivers certainty. Centers for Disease Control and Prevention data sets tell us how far apart we should stand, how long we may congregate and how well our community is doing. *The Last of Us Part II* has been shaped by prevailing cultural practices and, in return, provides a virtual field for developing new survival skills, new ways of seeing and listening and new navigational tools; it serves as a cautionary tale for being mindful of the presence of others, and an anxious parable about new ways of being in the world.

The evolution of artificial intelligence is both about developments of the underlying technology and its applied material context. From *Uncharted* to *The Last of Us*, Naughty Dog has evolved its systems, but the game narratives and their play contexts have also evolved. That one game has its roots in action adventure and the other in survival horror is significant, in terms of the emotional valence of their respective intelligence systems, even if these systems have a common programming lineage and a shared lineage in their play dynamics. Perhaps I have under-emphasized the storylines of *The Last of Us* and its sequel by over-emphasizing their common code work. As the franchise moves to HBO, that code work will inevitably be erased in the interest of episodic storytelling and by the nature of its new medium: television. This leads us to consider the distinctive value of AI in post-apocalyptic fiction and the distinctive nature of pandemic culture more broadly as it is expressed in playable media. What gets lost in hermetic dramatizations of the apocalypse? I believe that the answer lies in the qualities and activities of feedback, in exploration and discovery and in the balance between technical affordance (what a game can do) and accountability (what a player chooses to do). In *The Last of Us* and its sequel, you can craft Molotov cocktails and throw them at non-player humans, but do you want to hear them scream in pain? There is an inescapable cruelty in the two games, as they mechanize and animate both an unforgiving environment and a decided amount of human suffering. In both games, this is captured in the connective thread between narrativity, visuality and expressive intelligence, where the technical work of AI, pathfinding, for example, is often realized as a form of emotive responsiveness that illustrates both playable and non-player characters possess a heightened awareness of the world. I draw attention to the situated awareness of artificial intelligence across these games because it grounds our understanding of pandemic work more broadly, where survival is not a solo pursuit but an act of citizenship. To survive through *The Last of Us* and its sequel requires being attentive to multiple bodies, not simply as visual signposts of one's own progress but as responsive actors that can exhibit their own agency through the imprint of artificial intelligence.

4

NEW PLATFORM INDUSTRIES

Artificial Intelligence, Biometrics and Connected Fitness

The concept of playable media extends beyond the video game industry and the basic principles of gamification to a number of other industrial vectors that have expanded their reach through relational computation and the socialization of artificial intelligence. This chapter considers the turn toward AI in the home fitness industry (a movement reignited by pandemic uncertainty) and maps the contemporaneous emergence of several interactive ecosystems—connected fitness machines and systems that include Mirror, NordicTrack, Peloton, Tempo and Tonal—and the accelerated growth of interactive personal training and integrated fitness, media and technology companies. These ecosystems share a number of formal and structural attributes; they are all built from computational media systems, physical machines tethered to proprietary hardware, software, and live and recorded media (more broadly framed as experiential content). At the same time, these ecosystems share a number of transformative technosocial attributes, fostering community, reconstituting screen culture, and furthering the reach of programming into everyday life.

AI is transforming the healthcare field, where deep learning can lead to predictive diagnostics as intelligent agents sort through inputs and inform us how to act through data, and where health-related implementations of biometrics have yielded positive results and bolstered the public trust in biometric technologies. There are of course myriad reasons for collecting personal data, and not all biometric data aggregation benefits the public interest. Performance algorithms and risk assessment algorithms draw from a larger set of demographic indices and variables; and whereas pandemic modeling produced a series of baseline public health measures, the responsive algorithms of home fitness systems yield only a private benefit, for privileged consumers and for the industries that serve them and aggregate their data.

DOI: 10.4324/9781003225072-4

Life hacks, those shortcuts designed to increase personal productivity and efficiency, borrow from the vernacular of computing and the scripted shortcuts used by information technologists to streamline and accelerate their workflows. These emergent ecosystems are driven by the very mutability of artificial intelligence, activated as a network of intelligent agents that, in the interest of self-mastery, tie demographic data to biometric data and relay these values through a number of interlinked screen(ing) technologies: fitness machines, studio monitors and wearable biometric devices. These technologies give form to the quantified self and are aligned with a number of connected and predictive health devices, such as the Kinsa thermometer, which leverages AI as part of a sophisticated illness tracking system. These AI ecosystems run from consumer research to biofeedback; they market to and reinforce behaviors, take advantage of our freedom in order to harness richer data, learn from us, and work diligently to customize and strengthen their bond with us. These information loops lead to radical efficiencies in manufacturing, to the production of products and services organized around convenience and to the formation of a body of biometric knowledge that can be used to refine each system. Jennifer Whitson (2014) notes, "The unifying methodology of QS [the quantified self] is data collection, followed by visualization of these data and cross-referencing, in order to discover correlations and modify behavior" (346). The deep relations we build with digital fitness systems are grounded in the principles of visualization; in every case, but with unique interfaces, they make data visible, often by translating it into more familiar iconographic metaphors for such measures as time, distance and energy. Artificial intelligence increases the speed, precision and volume of data collection, and connected fitness machines tether our bodies to automated data scraping. As with procedurally generated games, these systems render space visible as we move through it; they provide real-time visualization (though not necessarily in three dimensions), as they track our progress, and AI is one of the principal elements used to translate what is a continuum of data into an understandable, goal-oriented language of signs. Using a gamification design model, these systems take a large and ever-expanding database and transform it into something users can quickly understand, communicate with and build community (Whitson, 2014).

Most athletic companies have turned to AI to strengthen brand loyalty while strategically building out their portfolios to lessen their dependencies on other technology firms. In 2018, Nike acquired Invertex, a leading computer vision firm based in Israel. This acquisition came one month after Nike's purchase of Zodiac, a consumer data and analytics firm. Nike has a history in the fitness technology marketplace, and although it has discontinued its FuelBand, an activity tracker that launched in 2012, it still has a strong foothold in the wearable application space. Peloton Interactive, which also launched in 2012, has been aggressive in its acquisitions. Since late 2020, the company has purchased at least four technology holdings—Aiqudo, Atlas Wearables, Otari and Peerfit—adding voice assistance, smart watches, interactive workout mats and a digital health platform to its portfolio. Each of these acquisitions has deepened the owner's share in AI research, and has allowed each company to leverage that expertise to build a dedicated ecosystem. Proprietary algorithms hold multiple value propositions within each of these ecosystems and tie them and their data together.

The operation of an information network—the flow of data from cloud-based systems to physical systems and wearable technologies—is multinodal in nature, a protracted communicative act that produces a number of interrelated transmediations (as data move across and bind together discrete media forms) that, writ large, shape data into a narrative that is grounded in a context and realized through structures that can produce agency: walk further, stand longer, train harder and breathe deeper. AI-powered fitness solutions have been largely recombinatory, adding wearable devices to existing fitness equipment, or upscaling conventional fitness machines with integrated biometrics, but the market has also seen innovation in the personal fitness industry, with AI-powered devices such as Mirror (acquired by Lululemon in July 2020), a large-screen display and personal training system that uses camera-based performance measurement; throughout the pandemic, these innovations have pushed into the home market, adding new intelligence agents to an already dense thicket of networkable smart home devices.

The pandemic has accelerated the growth of virtual training services and smart home gyms, and as popular brick-and-mortar fitness studios have temporarily closed their doors, closed entirely or imposed public safety provisions to comply with state-level mandates, downloads of health and fitness apps have been on the rise; many subscriber-based streaming services have experienced dramatic increases in consumer demand. A report issued by the World Economic Forum notes that health and fitness app downloads increased by 46% worldwide between the first and second quarters of 2020 (Ang, 2020). Within this marketplace, connected fitness equipment has become increasingly popular, as a key element of maintained wellness and sociability. The global home fitness equipment market experienced over \$1 billion in growth from 2020 to 2021 (Research and Markets, 2021) and led to unprecedented private investment with a number of key stakeholders vying for market share. Throughout the COVID-19 pandemic, these systems have eroded the market share held by retail gyms; data collected by the International Health, Racquet & Sportsclub Association (IHRSA), the fitness industry trade association, indicate that as of July 2021, 22% of health and fitness clubs and studios permanently closed their doors (IHRSA, 2021).

AI can process the vast amount of data from wearables, sensors, cameras and other connected devices, and the values associated with the home fitness market are just as much about functional hardware, as they are about the ease of the user interface and the ability of algorithms to aggregate and optimize the end-user experience, providing responsive feedback that can account for such variables as performance level, age, gender and body weight to target specific goals; these algorithms provide a significant return on investment to platform manufacturers, yielding robust inventories of consumer data. Companies are launching differentiated products, with distinct digital platforms, to solidify their positions and capture larger market sectors.

The digital home fitness market tracks with the rise of the wearable tech market which includes fitness trackers and smart watches. The Apple Watch, first introduced in 2015, holds the top spot in related market share, but is joined by several competitors that include FitBit, while it has eclipsed others such as the now defunct Jawbone, and has been shored up by early entrants such as Nike within the third-party app space. Throughout its history, the Apple iPhone has been associated with a number of health metrics, including a partnership launched by Nike and Apple in 2006 that paired the iPhone with the Nike+, a wireless in-shoe accelerometer. The iPhone, among many other devices, signals the ongoing deployment of communicative tools as personal technobiographic agents that also encourage consumers to share their otherwise personal data networks (Freedman, 2011). To court these consumers, Peloton and its competitors have built third-party integrations for popular wearable technologies.

The groundwork for home fitness was laid decades earlier and linked to a number of key technological developments. The post-war rise of television and the emergence of a suburban middle class created a market and medium for fitness gurus such as Jack LaLanne, whose show debuted in 1951 and was picked up for national syndication by the end of the same decade. Expanded sales of videocassette recorders in the 1980s created a space for fitness personalities such as Jane Fonda, who released her first workout tape in 1982. Nintendo debuted the Wii in 2006, in response to Microsoft's Xbox 360 and Sony's PlayStation 3, and followed up one year later with the Wii Fit and Balance Board, complementary software and hardware components that introduced exercise games (yoga, strength training, aerobics and balance) to the Wii platform. Wii's fitness-related games encourage a playful approach to biometrics; players are invited to take a body test, a gamified measure of body mass index and balance control that also generates a corresponding Wii Fit Age (an estimate of biological age based on fitness performance). The Wii's biometric outputs have questionable scientific value, and the company's playful engagement with the serious matter of body type and image came under fire by concerned parents. In 2008, the company asserted that its BMI assessment was only designed for adults and offered an apology to its consumers: "Nintendo would like to apologize to any customers offended by the in-game terminology used to classify a player's current BMI status, as part of the BMI measurement system integrated into Wii Fit" (Daily Mail, 2008). Along this timeline, which cuts across several mediums and genres, there has been a consistent market for home fitness equipment and an increasing emphasis on consumer-facing biometrics; the introduction of home workout programs significantly expanded the home gym movement, which gained traction throughout the 1980s and created a wider market for consumer-grade treadmills, bikes and elliptical equipment, a number of competing weight training systems such as Bowflex and Soloflex, and a number of unique methods for measuring and tracking health and wellness.

The Cultural Value of Algorithms

Connected fitness systems are an extension of an already evolved IoT built on the legacies of ubiquitous computing and leveraged from the widespread adoption of consumer-grade biofeedback instruments. Algorithms have come to dominate most sectors of our economy; they are everywhere. What has evolved since the arrival of portable computing is the relative convenience and familiarity of these machines, and the degree of trust we have placed in computational systems— systems that are wrapped as well-designed objects to deliver an extensive portfolio of applications and services. Within these systems, algorithms are performing significant cultural work, providing practical solutions, realizing efficiencies and shaping our lifestyles, or more succinctly, as Ed Finn (2017) proposes, working "as pieces of quotidian technical magic" (16) that monitor and facilitate our behaviors. We rule each day with algorithmic certainty, but perhaps more starkly, we have adapted to the logic of algorithms without necessarily fully understanding their syntax or command structure; we have learned over time how to frame questions to generate responses and move with greater speed from problem to solution through a series of tacit negotiations with our knowledge machines (Gillespie, 2014). It is therefore not surprising that in our quest for greater control over our bodies, in our desire for results, we have embraced the work of algorithms.

By pairing proprietary algorithms with protected data sets, developers can create particular and very localized (and readily monetized) logic models. The Peloton bike gathers our inputs during each ride before algorithmically providing our results or output. In Peloton's metric system, cadence (rotations per minute) and resistance (the level of difficulty set by turning the bike's iconic red resistance knob) are used to determine output (measured in Watts): information is collected, algorithms are applied and results are generated. Over time, these records of each ride, this catalogue of activities, becomes a richer data set; the cyclist becomes known by the machine in ways that are more personalized than generalized demographic variables such as age, gender, weight and height. Each subscriber's data set is engineered by database design and management to be legible, to conform to a graphical user interface that has both an internal and an external logic that gives it expressive power (Gillespie, 2014). The Peloton screen has powerful semantic dimensions that are complex expressions of a networked database architecture. That underlying architecture is, necessarily, formalized and organized to effectively communicate with the consumer. Tarleton Gillespie suggests, "For users, algorithms and databases are conceptually conjoined: users typically treat them as a single, working apparatus" (Gillespie, 2014, 169). Biometric tracking is, on the surface, a seamless operation, and it masks the privilege of being willfully monitored and warehoused as part of a larger data store. This is why it is so critical to conjoin social research on media and technology with research in the computer sciences, and to move beyond media texts and objects to consider the industries that produce them—industries that are deeply vertically integrated and populated by programmers, designers, hardware developers, consumer analysts and other market sector experts and developers. Algorithms become meaningful when they become material, and it is just as important to consider the materiality of playable media, the immaterial architectures of their programming and networks, the industries that produce them, and the consumers who activate them. The circuit of relations formed by these elements is a communicative and meticulously designed data loop, where the drive of industry, the formulation of patented machines and proprietary programming and algorithms is conjoined to customer data. The material technologies of play are bound to symbolic content; the technical and social conceits of playable media are interwoven. Responsive fitness devices are media and information systems that encode and organize work, and are often taken as satisfactorily direct and truthful representations of the world (Star, 1999) despite their indexical nature. Turn the red resistance knob on the Peloton bike to the right and watch the tension increase in logical numerical increments; the work of turning the screw, and the overall build of the frame, flywheel and pedals, is translated into a scalable and correlative digital value that can also be represented on screen. There is a connection between mechanical and digital labor, and between physical and virtual (visual) outputs.

The concept of infrastructure is a useful lynchpin between media studies and science and technology studies, and provides a unifying axis for organizing an otherwise messy taxonomy of distinct playable architectures. The algorithms that drive playable media fit within this model for understanding infrastructure in that they are situated architectures—organized code embedded within a localized development framework (the production pipeline) and encased in a series of media products that aggregate data from otherwise distributed information

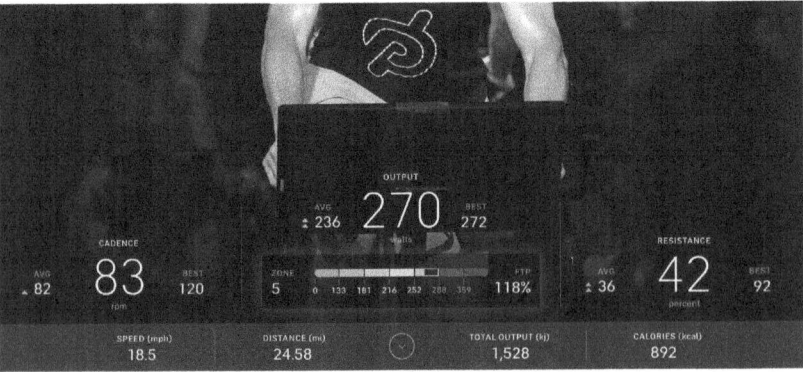

FIGURE 4.1 The Peloton Bike touchscreen displays an array of functional metrics, including cadence, resistance, output and power zones.

Source: https://blog.onepeloton.com/power-zone-training

networks (reaching across servers, cloud-based systems and local devices). Susan Leigh Star suggests infrastructure "is by definition invisible, part of the background for other kinds of work" (Star, 1999, 380). The information infrastructure of a large-scale technical system is not simply a neutral conduit, and when laid as the groundwork for a networked platform with integrated software and hardware components we might consider what, exactly, it is doing and what organizational problems, whether technical or social, it is solving (Star, 1999). Over time, the Peloton machine begins to know its user and can drive suggestions to the user with machine learning. To this end, the company's engineering team developed an unobtrusive recommender system:

> We picked a simple set of heuristics, which involved accounting for the Member's most frequent duration/type of workout and their most frequently used instructor. We computed recommendations for all Members offline, cached them and served them up when requested. Two factors drove this decision: Recommendations could be refreshed at a daily cadence without degrading the user experience; we were averse to querying workout histories in real time from our transactional database before we had a chance to apply these heuristics.
>
> *(Banerjee et al., 2021a)*

User and content are bound to each other through a chain of computation, and a series of Amazon Web Services (Glue, Lambda and DynamoDB, for data management and execution); and while the original recommender system was built with Amazon Personalize (a managed machine-learning service), in its final form, Peloton engineers designed and deployed an in-house machine-learning model (Banerjee et al., 2021a). While recommendation algorithms are designed to strengthen user engagement, they remain aligned with a company's business goals, which are often tied to promoting new features, services and content. Peloton engineers have integrated both values in its algorithm:

> We achieve these two goals by utilizing a combination of rankers and filters. Ranking models help order the universe of content for each user according to their preferences. Filters take a slice of this ordered content to generate a shelf of content with a reason for suggesting it. Explainability is heavily linked to business goals, while ranking is linked to engagement goals.
>
> *(Banerjee et al., 2021b, 575)*

While users arrive at the platform with a clear intent, and with particular affinities (to certain instructors and classes), Peloton's algorithm is able to introduce the user to its expanded library of content, and selectively promote new partnerships: "Recommendations serve as a mechanism to encourage them to try something outside of this comfort zone, which both increases the breadth of a user's engagement with the platform and leads to a more holistic workout routine" (Banerjee et al., 2021b, 575). The organizational problem at work in Peloton's digital infrastructure is intimately tied to its content library, as a media filter, and serves both technical and social functions.

Science, Technology and Media Studies

Where science and technology studies meet media studies, we can productively examine the experienced social relations of material media practices that are undergirded by software. From the perspective of science and technology studies, playable wellness media are sociotechnical assemblages; their technology and materiality are co-constitutive, mutually shaping each other through processes of translation and inscription; these fabricated ecosystems are integrated constructs of hardware and software, and are built by industrial designers, user experience and interface designers, programmers and subject matter experts to parse biometric data, stream it to an interface and make the body knowable. Yet these assemblages are not simply the sum of their technologies. Fitness machines are wedded to computational components and to networked data, but they are also bound to larger structures. Peloton is a global lifestyle brand, a provider of original scripted and cross-platform content, a talent agency with a large cast of athletic actors and a number of high-profile entertainment industry partnerships (that include collaborations with David Bowie, Taylor Swift and Beyoncé). It has broadcast studio holdings in London and New York that employ tightly integrated crews to oversee physical production. The company has been featured in *The Hollywood Reporter*, where it was aptly described as "an old Hollywood studio, creating star instructors who are on multiyear contracts, producing all its own content and controlling its distribution" (Rubin, 2021). While Peloton may be leaning heavily on technology to capture a significant share of the fitness industry, it is also taking over a portion of the broadcast production spectrum and dependent on a number of key physical assets and personnel. Peloton is drawing from the evolving landscape of studio-based production, and its classes are carefully scripted, well-choreographed videographic exercises between humans and machines:

> Inside of the Peloton studio in Chelsea, NYC, the instructor leading the class is captured by four Panasonic AW-HE130 pan/tilt/zoom cameras, one mounted on a permanently installed Telemetrics TeleGlide track system hung from ceiling that helps it capture the instructor with smooth dolly-style camera motions that cover 180 degrees around the main instructor stage.
>
> *(Telemetrics, 2017)*

By adopting robotic camera control and tracking technology, Peloton's segment producers are able to consistently conceal the apparatus of production and hold onto the qualities of immersion, while creating a dynamic interplay between their instructors and their members; the custom bike is the only apparatus on display. While a data-centric machine-based system (of robotic cameras and networked exercise equipment) sits at the heart of Peloton's enterprise as the primary structural support for its membership base, as a grand-scale producer of playable media Peloton stands as a media, fitness and technology company, and its output highlights the increasing confluence of multiple industrial ecosystems—broadcasting, music, industrial design and wellness—that have all been recontoured by software, and the ability of capital to speed along that change. Within this convergent ecosystem, Peloton and other vertically integrated media and technology companies are able to neutralize (a stronger sentiment than mere appropriation) the objects they add to their respective portfolios, conform them to the logic of their programming, and attach new values that harmonize these acquisitions with a brand image. Peloton's rights to the full David Bowie catalogue and the introduction of the artist's songs into its interactive fitness classes may create a level of discord, but that rupture can be quickly sutured; each track gets imprinted with the exercise tablet's graphical user interface, attached to signature metrics, subdivided into performance zones, and divorced from its counter-culture origins. The Peloton brand is largely defined by the complex sum of content, data, software and hardware, but Peloton and its competitors are significantly expanding screen culture by assimilating aural culture and culture writ large.

Digital fitness is part of a larger industrial complex focused on measuring and regulating health and wellness and bound to a belief in computation. Diet plans are grounded in calorie counts, those with meal plans are grounded in macronutrient ratios, and those with therapeutic elements are grounded in numeric understandings of persistence and well-being. Applications such as Noom include many of these ingredients. Noom tracks food intake and exercise and links these to quantitative demographic data as well as qualitative experiential data that describe the user's lifestyle and goals; in turn, the application provides personalized feedback, through an algorithm, through coaching, and through other external supports. Applications such as Noom are extensions of earlier calorie counting systems, the centerpiece of many commercial weight loss programs, and add correlative values between body and lifestyle. As a whole, these semi-intelligent systems, either analog or digital, share several basic principles of self-regulation, including self-monitoring, goal setting and self-control (Chew et al., 2021), but they lack the needed syntax to parse complex social dynamics. The varied applications of artificial intelligence, for real-time and predictive analytics, are quite efficient at mapping personalized goals along a temporal axis, but do little to reveal how, precisely, qualitative behavioral frameworks can be realized as a simple graph.

Beyond its technical function, code has been used as a critical construct to speak of structure, architecture and regulation, but its regulatory function has been consistently reframed as a positive attribute in software-dependent wellness media. Lawrence Lessig (2006) writes quite openly about the regulatory power of code, and about code as law. Code has functional rules and protocols, governing standards, and it flows from a set of necessary technical procedures for managing and distributing information. Eugene Thacker (2004) suggests that the protocols of networked communications have particular biopolitical and material dimensions, in the sense that networked bodies are inscribed and produced by informatics (xix). Connected fitness equipment is designed with the express purpose of self-regulation through observation, of discovering ourselves through an array of digital artifacts. We should recognize the relationship between code and control and its ongoing abuses, but we should also amend the arguments of Lessig and Thacker with a caveat; while connected fitness systems seem to confirm that when code is wedded to parent technologies that conceal its labor and its social and economic imperatives, it cannot be ideologically neutral, that these systems blur the boundaries between healthy self-regulation and harmful bodily governance, we need to acknowledge that individuals have unique and often playful relationships with technology. Not all technobiographic arrangements and practices are oppressive.

Playable media visualize data through their various enactments, and are typically aligned with the dynamics of embodiment. Science and technology studies have, over time, expanded beyond information technologies to consider media technologies as yet another set of situated cultural practices that have been influenced by the steady growth of computerization. Science and technology studies when paired with communication and media studies can help unravel some of the complexities of playable media systems and tease apart the layers of code, machine and human-centered engagement. We can parse the interface and machine layers of these systems from their core technologies, their code and their data stores (which manage various collections of data, including those harvested from interactants); and we can also consider the transmutability of biometric information. Each ecosystem has its own iconography, its own benchmarks and its own vocabulary, although these signs must be familiar and recognizable; output must be made meaningful to Peloton users in the same way that "splat points" (a measure of time spent in higher intensity training zones, which are in turn a measure of heart rate) must make sense to Orangetheory members. These insular vocabularies are a way of building community, benchmarking progress and motivating consumers; as these emergent terms develop into stable group-specific understandings and ways of sharing information, each user group gains the potential to imagine itself as a community and each individual can be positively reinforced through very real practices of affiliation (Baym, 1998).

The social contexts of science and technology have been furthered by their ongoing influence on screen culture; beyond its algorithmic logic and its function as infrastructure, data are a foundation for user interface development where it also performs symbolic work as a formal sign of user experience. For quite some time, screens have been a mainstay of public and private space, transforming public settings and shaping the dynamics of home life (McCarthy, 2001). Mobile devices and practices have fully normalized the migratory patterns of screen culture. Yet, the rhetoric of spectatorship, of place and of practice has been further complicated by the data-centric turn in American life and culture. Not only have we adapted screen culture to new social rhythms, remapped screen real estate to accommodate complex and multilayered data streams, bound it to monitoring both personal and communal quality of life indices, tied it to qualitative and quantitative assessments, and used it to regulate homeostatic and environmental conditions, we have also expanded its role beyond traditional spectatorship and ingrained it as a natural interlocutor, as a trustworthy agent, and as an active mediator, translator and partner. Screens have been tasked with carrying out an increasingly broad and dynamic range of social and institutional functions—a range that has expanded far beyond the repurposing of the television set to play console-based video games, first with the commercial release of the Magnavox Odyssey in 1972, and then with the entrance of Atari, Coleco, Mattel and Nintendo into the home console market across the same decade. Interactive home fitness systems embody a series of remediations beyond the screen, and certainly beyond the television set: the stationary bicycle, the mirror, the treadmill made digital, responsive and interactive. iFIT, a system first launched in 1989, is a curious bridge technology between early model ProForm treadmills, NordicTrack elliptical cross-trainers and consumer-grade VHS players; these home fitness machines effectively handed device control over to a series of workout tapes and

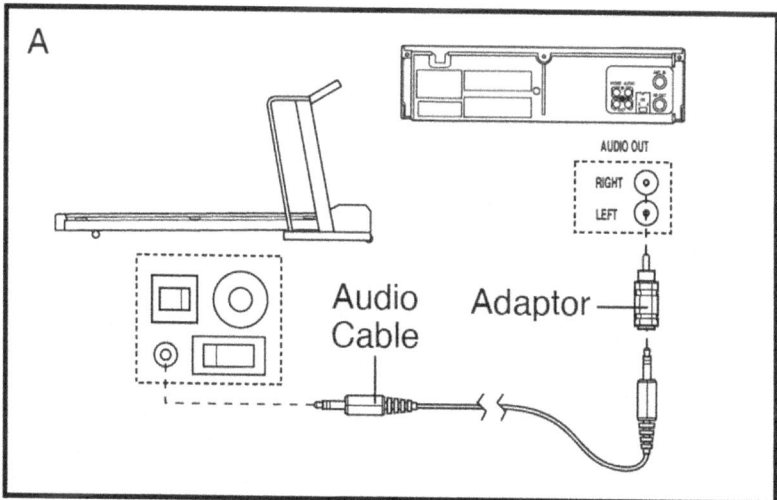

FIGURE 4.2   The user's manual for the ProForm 10.5 QM includes instructions for connecting the treadmill to a VCR using an audio cable for automatic device control.

Source: Copyright 2003. ICON Health and Fitness

through an embedded audio signal could automatically adjust their speed, incline and resistance to follow along with a pre-recorded instructor.

The textual processes of looking, engaging and performing vary even across this abbreviated history of the expanded screen. Yet I am suggesting that the body serves as an important arbitrator between media and technology; it stands as a site that can be activated to fuse together discrete forms of representation, and it stands as a site that can negotiate between technology and society. This history of bodies and technologies is undoubtedly imprecise. How do we track successive innovations in digital fitness when these innovations intersect with other institutional histories—those of computation, media and human physiology—and with broader changes in social and cultural practice? We can certainly describe these socio-technical networks in some detail, how these systems are constructed and how they function, but explaining their evolution is a grander challenge. Idealistically, Bruno Latour (1990) suggests the explanation may emerge "once the description is saturated" (129). This is why it is important to understand the information layer of these systems—to understand what, precisely, is being communicated and to whom—at the same time that we consider their forms and functions as embodied media. Digital fitness systems are heterogeneous assemblages of data and machines, informed by several distinct institutional histories and built from a number of unique material artifacts, each with its own legacy; these technologies are artificial constructs, financed, researched, developed and assembled to meet a series of culturally validated goals, dependent on their actors, and they echo Latour's sweeping invocation that "technology is society made durable" (1990). Digital fitness systems are points of industrial convergence that signal their context.

The Business of Socialized Data

These systems, regardless of their respective footprints, are all tied to a network. Each line of connected fitness equipment is linked to a subscription-based service that offers access to a catalogue of fitness classes and a method for archiving biometric data. The Peloton Bike and Peloton Tread feature large HD screens to stream classes, access a library of pre-recorded content, visualize biometric data, display leaderboards and provide live feeds to connect users through virtual high fives and social media tags. The Peloton Guide, a more recent addition to the company's portfolio, utilizes a full-body movement-tracking camera and purports to transform any room with a television into a strength-training gym. The Guide more fully realizes Peloton's investment in machine learning by tracking and learning from each interaction as it monitors the participant's movement patterns. Peloton follows a vertically integrated business model of proprietary hardware, software and media, what the company refers to as a technology stack of its physical equipment, its gamified application software and its streaming video content (Peloton, 2020). Within this stack are a number of key infrastructure components, with the bike and treadmill portfolio being complemented by integrated touch-screen technology and speakers, the client-side software bundled to Python Cloud Services, and the live and recorded television content dependent on the physical development of its private recording studios, and of course the labor of its fitness instructors. The company also prides itself on its social structures, its membership community, which is connected through a number of common social media threads that include Facebook and Instagram, and the managed accounts of its personal trainers; trainers can proactively push out motivational video messages at key milestones to individual users and can shout out to users to acknowledge their achievements during live classes.

Performance measurement and biosocial feedback are critical metrics of progress: the social dynamics of the workout class, the leaderboards ranked by member outputs, the virtual high fives in the personal feed, and the aggregated data sets that monitor and reward participation all function as signposts of the communal work of Peloton's platform. The promotional video for Lululemon's Mirror addresses its members as a community: "See yourself, and know you're not alone. Because this is not just a mirror. It's a reflection of an unstoppable community." (Lululemon, 2020). Mirror is a fully realized technobiographic system (a system that writes the self through technology); it captures and reflects each member's physical body, performs symbolic work through its workout statistics and reads the individual as part of an aggregate across its membership network. Mirror reflects and projects simultaneously; during each workout session, it superimposes the trainer's image over the participant's image and scales these to create perspective. Through its layered visuals, Mirror demonstrates its algorithmic power.

Mirror and its competitor systems objectify a more widespread data dependency that has enveloped our culture and our economy. These device-centered ecosystems are changing the aesthetic vocabularies of the traditional media forms that they all draw from. The form and function of personal training videos are changing the landscape of broadcast production and creating a new studio economy based on altered notions of seriality and progression, and distinct timetables that accommodate the flows of live, on-demand and archived content. More broadly, the available inventory of streaming content is also becoming more subdivided and hierarchical, adapting to new revenue models in a market saturated with subscription-based services.

The vertically integrated media enterprises that have been realized and are owned by Peloton and its competitors are sites of privilege. Customers with a household income of more than $50,000 make up 92% of Peloton bike owners (Peloton, 2020). These connected, screen-enhanced machine-based ecosystems are part of a larger circuit of body culture media that has expanded the parameters of the fitness industry and allowed a number of tech entrepreneurs to add health and wellness to their investment portfolios. Peloton, Mirror, Tonal and Tempo all manufacture exercise equipment, produce media and develop technology. Mirror is an LCD (liquid crystal display) mirror with a quad core processor, a front-facing camera and a stereo sound system. Tonal is a wall-mounted digital weight machine with a glass touch-screen interface and adjustable steel arms. Tempo Studio is a free-standing weight storage cabinet with a top-mounted screen.

Peloton, Mirror, Tonal and Tempo all connect to live and on-demand content within their respective networks, and while these service providers speak loudly about their personalized training, they signal their difference through their hardware, their display technologies and their intelligence operations. However, artificial intelligence resides in more than one place within these systems. AI is a term that readily suggests that each device is intuitive, responsive and customizable, and can provide personalized feedback and recommendations; these systems are designed to learn from their actors. However, the concept of intelligence is used to describe a broad-ranging inventory of agentive features. Loosely defined, it suggests a range of responsive practices that emerge from a number of recombinatory hardware and software solutions; and it suggests a range of computational methods. Tonal promotes a patented digital weight system that tracks repetitions, sets, power, volume and range of motion, and makes these measurements readily accessible. Defining the mechanics of its hardware, Tonal describes its system as an electromagnetic resistance engine: "Using a computer chip doing millions of calculations per second, our digital weight system can mathematically model the physics of a traditional weight stack and digitally manipulate that electromagnetic field to make it feel like a weight stack" (Tonal, 2018). Tempo's embedded screen technology uses 3D sensors and AI to analyze the user's form and provide feedback during training sessions. Peloton's acquisition of Aiqudo suggests that the company may be adding natural-language voice commands to its systems, which has been the focus of Aiqudo application development.

These specific features point to deeper investments in artificial intelligence and machine learning within each company. In the popular press, AI has been broadly conceptualized as adding feedback-based personalization to an evolving digital fitness industry—deployed to learn from individual exercise and recovery patterns, to serve as a recommendation engine and to activate form-adjusting or corrective technology. Digital fitness has been formalized as an informatic loop that links software and hardware and collects personal data as part of system-level performance tracking, focused physical improvement and responsive training and engagement. To engage their member base, AI-enhanced fitness machines, applications and programs deploy machine intelligence for two primary purposes: assessment and guidance. Most of these systems promote an active understanding of their algorithms through a series of highly visual performance metrics: workout statistics and historical graphs that track progress and highlight milestones. As they trade in a common vernacular that can aptly describe human performance, most of these systems resituate AI in more familiar terms and within their own social contracts. Tonal calculates strength scores (and references physical weight stacks), while Peloton offers functional threshold power; each system has a localized vocabulary of benchmarks that serve as motivational journey maps, and all of them include a number of social features and challenges to keep their members engaged. The journey is not simply a qualitative term; it is a formal construct produced by programming. Peloton employs software engineers (a Journey Cloud team) to design and develop its "workout journeys." There is a syntax to the workout experience that also governs the personalized home screen of each bike and treadmill, and that clear syntax allows workout data to flow to third party platforms, connect to achievements and challenges, and create performance histories; the journey is a map that guides each exercise routine and marks progress, but it is also a coordinated series of paths that guide the quieter movements of consumer data.

Expanding Platform Studies

In June 2021, fitness equipment supplier ICON Health and Fitness changed its corporate name to iFIT Health and Fitness, adopting the name of its proprietary software platform. iFIT owns three distinct equipment brands (Freemotion, NordicTrack and ProForm) and has developed an integrated system of software, hardware and live and on-demand content for its commercial and residential clients. Connecting its users to the outside world, iFIT delivers a range of experiential content: "Trekking to Mt. Everest basecamp on a treadmill. Rowing the Thames on a rower. Cycling by Kyoto's temples. These are just some of the iconic experiences our unique experiential content emulates" (iFIT, 2021). In October 2021, iFIT announced partnerships with the Boston, Chicago, London and New York City marathons to provide additional content for its connected treadmills; and through a partnership with Google, users can create custom routes on Google Maps and build immersive training experiences that draw from real-terrain data and images. As with its competitors, iFIT technology is part of an integrated yet flexible platform; the common operating system that unites its experiential content with its interactive hardware also provides an iterative architecture that can expand the platform and pull the world into its orbit.

By following a model first developed within platform studies, we can refine our understanding of fitness-based technology companies, each of which offers up the promise of a comprehensive wellness platform, by shifting our attention away from their libraries of experiential content to consider the materiality of their technologies and the operations of their code. While we can understand platforms as well-designed and often sealed computational systems, we can also consider the possibility of their reactive nature, with people doing more than simply performing an underlying algorithm. Ian Bellomy (2017) proposes, "Human computation is experienced from the first person; the process has a phenomenological dimension to it." While people do perform through a system's algorithm, they also produce procedural artifacts that are signs of their experience, and that experience includes a fundamental awareness of instantiating those underlying processes (Bellomy, 2017).

As content delivery systems streaming live and on-demand classes and ancillary video and audio content, each of these machines is a playable media apparatus; and while they gamify fitness by purposefully stratifying the experience into a series of functional level designs, rankings and achievements, they pursue grander goals. By measuring and metering the body, they encourage particular forms of performance that are both structured (mimic the movements of the trainer, move within certain parameters) and exploratory (find your Functional Threshold Power and define your Power Zones). The relationship between body and image and between body and machine defy traditional game mechanics, and are more closely aligned with functional training than they are with the varied expressions of play *per se*. Here, computation provides an organizing frame, a way of decoding the body, which seems counter to the general impulses of artificial intelligence. Where the anxieties attached to AI reflect the notion that machines will become increasingly human, in the wellness industry, AI is a lynchpin for understanding the body. AI purposefully transforms us into machine-readable subjects, not to know and conquer us, but to help us understand ourselves—to become more efficient engines. This is part of a larger cultural turn toward "computationalism" (Golumbia, 2009)—an increasing dependency on and belief in the power of computation. However, this is also the product of the growth of network culture, the expansion of information systems, and the institutional, organizational and social formations borne out of distributed communication. In an effort to expand its customer base, and in a way that foregrounds the company's investment in software, Peloton has turned to more explicit forms of gamification with the 2022 release of *Lanebreak*, an interactive fitness game that remaps the bike's resistance knob into a game controller and replaces the instructor with a wheel-like avatar that can move across variable lanes of resistance along a virtual neon-colored bike path; the *Lanebreak* fitness application effectively translates fitness goals into a more traditional game-based rewards system of points and high scores. The application also signals the company's ability to shape the fundamental mechanics of its bike across multiple algorithmic axes, swap content and change the interface while holding onto several core concepts of mechanical translation, such as resistance and cadence; the application stretches but does not violate familiar patterns of interaction, and by doing so, it rewards the domain knowledge of its customer base.

Digital fitness machines and programs seem to have solved the mechanical translation problem raised in early computer intelligence research—a problem that was expressed quite forcefully during the early years of machine translation research in a series of correspondences between two pioneers of the field Warren Weaver and Norbert Wiener. The dialogue between Weaver and Wiener was grounded in a series of related concepts articulated by Claude Shannon in his landmark work on communication theory, first published in 1948. Together, Weaver, Wiener and Shannon outlined the technical and semantic problems associated with transmitting human communication, and laid much of the groundwork for the field of information theory. In his 1949 essay, "The Mathematics of Communication," Weaver repeatedly distinguishes between information and meaning, and summarizes the difference with the rather reductive (and from this vantage point, sexist) scenario:

> An engineering communication theory is just like a very proper and discreet girl at the telegraph office accepting your telegram. She pays no attention to the meaning, whether it be sad or joyous or embarrassing. But she must be prepared to deal intelligently with all messages that come to her desk.
>
> *(Weaver, 1949, 14)*

By drawing from information theory, we can see that the certain and central power of data-driven fitness systems is their ability to commingle the two variables of information and meaning—to make information meaningful, to channel the entropy associated with otherwise open communicative processes and relay our bodies back to us in no uncertain terms. Although they serve a unifying purpose, the sealed computational systems of the digital fitness industry have expansive frames. Each proprietary system exceeds what we might consider the more conventional limits of earlier hardware platforms (such as video game consoles), and for this reason, it is harder to pin down the embedded possibilities of their respective software frameworks; each operating system is attached to a large number of data holdings and is designed to be much more flexible. And while the form factor of the physical machine may be fixed, the screen is a more open signifier, its interface easily re-architected. While Peloton's financial outlook fell precipitously in early 2022, largely due to the overproduction of its hardware and a drop in consumer demand, the company's software systems provide a way out, a way of building new design problems, new challenges for its members, into an existing circuit of relations that links fixed analog functions (cycling, running) to the fluid experiences of computer programming. Fluctuating market conditions have had an adverse impact across the home fitness industry; in October 2021, iFIT Health and Fitness announced the company was postponing its planned initial public offering just one month after first announcing its plan to go public. The changes in demand that have rippled across the home fitness industry are likely tied to lifting COVID-19 restrictions in the United States and in other major markets, and the reopening of brick-and-mortar studios.

The material shifts of these platform industries, and their ability to pivot toward new opportunities, are largely dependent on their labor practices, which are in turn tied to realizing a value proposition built on three elements: proprietary software, experiential content and interactive hardware. Peloton employs a deep bench of talent to support its platform: content producers, user experience and interface designers, enterprise information technology specialists, software engineers, warehouse associates, financial analysts, hardware developers, marketing associates, membership managers, retail experts, product engineers and studio operators. Peloton has been actively recruiting new leadership in artificial intelligence, machine learning and computer vision to create new roadmaps for these technologies within its corporate portfolio. The company's employee base continues to pivot with its platform. This remains a stark reminder that the possibilities of a platform are tied to its labor and driven by a particular approach to human capital; as it develops and delivers new services to its platform, a company must change the operations of its code.

David Golumbia (2009) suggests that the most dominant and worrisome use of computer technologies is that they will increasingly striate our cultural existence, while they operate as private, proprietary and patented formulations that put them beyond public scrutiny. Digital fitness developers continue to refine the communicative relays of their systems, learn from our behaviors and create new dependencies that keep us folded into their membership. Behind every new strategy is an opportunity. Alongside shifts in human (platform) capital, closely held patents, which are the product of that knowledge capital, are driving the digital fitness wars. Several were filed as early as 1989; these are focused broadly on control mechanisms for a series of exercise apparatuses and highlight the conceptual and (possible) practical interplay between computation, content, control signals and console development. United States Patent 6,312,363 B1 filed in 2001 on behalf of ICON Health and Fitness titled "Systems and Methods for Providing an Improved Exercise Device with Motivational Programming," summarizes the history of related filings: an exercise machine coupled to a VCR, an exercise machine coupled to a video game device, an exercise machine coupled to a communication module, and an exercise machine coupled to a remote evaluation module (Watterson et al., 2001).

ICON Health and Fitness and Peloton Interactive have been engaged in a number of patent infringement lawsuits since 2016 that include disputes over their respective hardware and technology holdings amid their respective pursuits of real-time trainer-based interactive systems. ICON holds a United States Patent for "Coordinated Weight Selection," described as:

> A free weight assembly includes a cradle, an aerobic exercise element that is movable with respect to the cradle during the performance of an exercise, at least one free weight removable from the cradle, an input in communication with a processor that determines a time to instruct a user to remove the free weight, and an indicator that activates when the time to remove the free weight arrives.
>
> *(Watterson et al., 2020)*

At issue in this patent lawsuit, which provides a rather detailed description of an interactive weight training system, are both the hardware (the Peloton bike has a rear-mounted weight rack) and the graphical user interface. During hybrid interval and weight training sessions, the Peloton screen displays a linear timeline with relevant icons that let the user know when to lift weights and when to ride the bike. ICON has suggested that this configuration infringes on the company's patent which describes and claims "technology to automatically control an exercise device in conjunction with an interactive integrated weight system" (ICON v. Peloton, 2021). Both ICON and Peloton claim to have unique hardware and software integrations, with proprietary features built into their respective tablets and as part of each screen's graphical user interface. United States Patent 9,174,085 B2 issued in 2015 to (Peloton co-founder and former Chief Executive Officer) John Foley titled "Exercise System and Method" describes in detail the circuit of relations between Peloton's remotely delivered classes, its central server and its proprietary bike and display screen (Foley et al., 2015); while the company's lawsuit against ICON Health and Fitness suggests ICON machines with iFIT functionality infringe on several elements of Peloton's intellectual property conveyed in the 2015 patent, including its leaderboard technology. Peloton's patent portfolio has come under scrutiny, partly because of the range of battles over intellectual property rights with ICON and other competitors that include Echelon and Flywheel; in 2020, the latter admitted wrongdoing and settled a patent-infringement claim first filed by Peloton in 2018 over its leaderboard technology (Peloton v. Flywheel, 2020). Peloton's assertions, argued in a series of patent filings and disputes over intellectual property, have cemented the company's brand name in a market driven by delivering responsive content and having the technological supports to do so.

In its "Exercise System and Method" patent filing, Peloton calls out the myriad ways its systems can gamify the user experience: competitions, badges, trophies (Foley et al., 2015). These reward structures are intended to incentivize goal setting, improve performance and encourage exploration. Despite the patent filing, these structures are not at all unique; rather, they are attached to a number of recognized conventions for effective user interface design, among them providing regular feedback and communicating through familiar signs and concepts that allow the user to understand what is happening within the system (Nielsen, 1994) and where they are in a particular process.

While they may seem beyond the purview of platform studies, patents are a critical reminder that artificial intelligence is not simply a thought experiment, but an instrument that can be lodged in material practices and produce tangible results. Patents are also a reminder of the general attachment of productive values to artificial intelligence in these contexts, at least from a purely legal perspective, as articles that act contrary to the public good are generally not considered patentable. Undoubtedly, we should reattach these patents to a more rigorous analysis that considers how they function, once developed, as social systems (Crawford and Calo, 2016), and how their impact may ripple outward through society in unforeseen, deleterious or inequitable ways that reify existing power structures.

Each of these digital fitness systems must attach itself to a dominant pattern language that allows it to capture the public interest (by effectively communicating the what, why and how of the invention) before it can move public discourse. Collectively, they embody the general qualities that are celebrated in an industry geared toward efficiency, to sustainable and measurable results, to the production of self-empowered (though machine-dependent) individuals. As we play and experiment within these systems, we discover ourselves through their data; informatic loops operate within our pleasure principle by providing regular signs of our effort and progress. Within these systems, we begin to establish new biometric literacies and in turn develop new understandings of our bodies; our labor is translated into a material measure of our output and reflected back to us on screen. These forceful signs of progress are difficult to abandon, and so we begin to place our faith in yet another algorithmic system.

The interface, through its symbolic constructs, registers our ability to effect change, but to realize our goals, we need to be expressly open with our data. Machine algorithms depend on the open exchange of information, and the commercial rhetoric of computation assures us this is a necessary step in charting our path to wellness. Machine-learning models perform their predictive analytics by drawing inferences from our personal health data, but these are imperfect impersonal agents functioning through linear, random forest or decision tree regressions, support vector machines and artificial neural networks (all discrete forms of machine learning), and they are only presumed to have some clinical utility (Chew et al., 2021). In fact, these are not clinical systems; they are first and foremost branded ecosystems of patented machines and proprietary applications, industrial design and fabrication, broadcast studios and streaming content, retail showrooms, gear and apparel, personnel, members and aggregated consumer data. These systems are driven by their algorithms—algorithms that shape social media, elevate sales, act as recommendation engines, and illustrate personal effort and results, and are purposefully expressed through their respective interfaces. Machine learning serves as the connective tissue; the platform is the product, and these competing fitness brands all look to optimize their algorithms, make their data cleaner and easier to analyze, and know their consumers. Health and wellness are part of the brand and are attached to the collective identity of the membership.

The algorithms of the digital fitness industry mirror the algorithms, both in form and function, of the more pervasive computational processes that govern our culture and institutions; they aggregate personal data and use a number of analytics engines to parse that data to set forward a series of recommendations (Finn, 2017); yet their interfaces reveal very little about their processing work and the underlying event-driven architectures that are responding in real time to what users are doing. Programmable fitness machines are only helpful (and thus profitable) if they are both precise and prescriptive; we rely on their data, their real-time indicators and visual histories to track our progress, and we assume that they are relaying back to us an accurate accounting of our bodies. These machines strive to individuate us, and the admixture of machine programming and human personal training works to that end, but at the same time, the governing logic of these machines, tied principally to their programming, also integrates us, not simply as a member on a leaderboard and as part of a community of like-minded members, but as stored information that can be analyzed and reported from a data warehouse and as real-time physiological data that can communicate within a platform. The Peloton bike and its head unit communicate with each other through a series of protocols that translate user movements into measures of cadence which are calculated against measures of resistance to produce measures of output; and these signals are processed with minimal latency to produce reliable and responsive signs of performance on the bike's attached tablet screen. Beneath its familiar armature, the Peloton bike is a data platform driven by a large number of integrated infrastructure solutions that also pull content from the company's central servers. The bike is designed to respond to and mediate the human body, and machine learning allows it to do so in context and over time (Kolko, 2010). By exploiting their respective algorithms, Peloton and its competitors have created new spaces for efficient real-time visualization, deepened our dependency on computation and tightened our attachment to the IoT.

These systems, however innovative they are in the particular, however stratified they have become through their patented machines and technologies, all capitalize on the inherent way that our culture uses data: to measure, organize, analyze, synthesize and act in the world (Greenfield, 2017). The connected machines of the home fitness industry measure our bodies to produce data and they organize, analyze, synthesize and relay that data in actionable form. The path to meaning from data to information is one that involves collection, calculation and visualization, and connects hardware, software, algorithms, interfaces and screen technologies. We naturally assume that the model of knowledge generated by connected AI-powered fitness devices is based on human knowledge—that the measures, however proprietary their values or icons, are grounded in the health sciences.

These distinct fitness solutions operate as platforms, as intermediating devices between data and information; they read and write the body. They record us and relay ourselves back to us as graphical (user interface) subjects. As platforms, they are not only conduits for data and information; they also shape information policy (Gillespie, 2010) and the more decisive fields of artificial intelligence and biometrics. These devices function in multiple ways as platforms, a term openly embraced by their developers; they are computational and physical structures, and they are figurative constructs (Gillespie, 2010) that offer a functional ground for health and wellness. For these subscription-based health and fitness technology companies, the term platform (which is used openly in marketing materials, press releases and investor reports) acts as a boundary marker for their systems and an entry point to content and community, and situates them as holistic enterprises that can be differentiated in the marketplace and protected in a court of law. The competition between them pushes each of them to pursue an ideal model of seamless interactivity and integrated subjectivity (providing complementary supports for physical fitness and mental health); yet if we focus too much on software and hardware, and on technology and design, which are indeed central to understanding how these systems situate their members, we risk losing sight of the labor that sits within them. The platform is also powered by content, and that content is built from the work of its trainers, segment producers, technical engineers, programmers and interactive media developers. Jamie Woodcock (2021) argues: "Platforms are a novel organizational form. They use digital technology to position the company (or 'platform') between the worker and consumer. This means they mediate the relationship between them" (6). These interactive systems are designed to sustain a series of assimilable dialogues: between the customer and the product, across members and trainers, between end-users and system analysts, and among the company and its shareholders. Within the digital fitness industry, these interplays are physical, emotional, functional and technological (Kolko, 2010) and they are articulated through expressive data and design solutions that conceal their own industrial methods.

Connected fitness machines have transformed our bodies and our homes; they have the potential to create what Michel Foucault (1986) describes as heterotopic spaces, "capable of juxtaposing in a single real place several spaces, several sites that are in themselves incompatible" (Foucault, 25). These physical systems are marked by artificial intelligence and made over by gamification; they function as both information spaces and as sites of goal-oriented play, and play is a central force in socializing these technologies (Ellerbrok, 2011). Connected fitness systems are designed to create behavioral change, and as they shape our bodies, they also (positively) shape our relationships to technology; we invite them into our homes and into our lives.

Connected fitness machines have emerged from a larger paradigm of quantitative self-analysis, but they have done so on their own terms. They operate within walled ecosystems with tight feedback loops that are supported by artificial intelligence and whose analytical signs are conveyed through assimilable real-time visualization. Gary Wolf, one of the co-founders of the Quantified Self project (along with Kevin Kelly), has suggested that "self-trackers seem eager to contribute to our knowledge about human life" (Wolf, 2009). Peloton and its competitors have intensified this appeal to community; each brand situates its individuated biometric markers as part of a collective. Artificial intelligence speeds this process along as it autonomously collects, organizes, aggregates and socializes our personal data; it runs scenarios, monitors our behaviors, shifts parameters, identifies peer groups, generates comparative analytics, makes recommendations and tells us how we can better ourselves. We place our trust in these proprietary algorithms because of our innate desire to be social, and because they trade in the familiar signs of playable media. Ariane Ellerbrok (2011) has tracked a similar evolution in the field of facial recognition, where the technology has been disassociated from its controversial origins in surveillance and attached to a sunnier inventory of playful AI applications, embedded in a suite of social media platforms where it can be imagined as something qualitatively new and different and where its code base can remain concealed; one such application, *Voilà AI Artist*, is discussed in Chapter 1. While serious technologies can be reconceptualized as playful technologies, they often continue to reinforce existing power relations, despite their apparently playful dispositions. This brings us to consider the broader power of playable media, beyond the localized and rather unified goal-oriented work of connected fitness machines, to examine how more expansive media practices utilize play as a framework to legitimize and socialize artificial intelligence within larger affiliate groups while continuing to obfuscate the serious work happening deep within their programming.

5

WORLDBUILDING AND DIGITAL TWINS

Beyond its application within bounded media objects and experiences, machine learning is driving new efforts in worldbuilding and is a key feature of exploratory AI ecosystems (those sites where data can be said to coalesce) that are driven by complex stakeholder networks; here, media and technology companies commingle with the defense industry, with industrial investments in transportation, urban planning and development, and with a list of other investors that are far too long to list. These stakeholders, whose interests are not always aligned, hold common investments in the architecture of media and information and the systems and resources of communication. For some, these frameworks drive product development; for others, they drive the reimagining of public and domestic space. Media and information networks are central to public sector deployments of AI, and while AI does not have an inherent or predetermined horizon, the fact that it is embodied and made material through these networks, through the machinery of connected data, suggests that it inevitably commingles with both ideology and industry on the path to taking on its various physical forms. AI is not immune to the influence of social, political and economic structures and the legacies of its historical lineage in computing and related intelligence fields. In the *Atlas of AI*, Kate Crawford (2021) puts a finer point on the matter of structural influence noting, "due to the capital required to build AI at scale and the ways of seeing that it optimizes AI systems are ultimately designed to serve existing dominant interests" (Crawford, 2021, 8). AI is both technical and social, in its apparatus and in how it uses data to produce and circulate meaning about the world. AI is generating a world view, expanding its mapping functions from objects to buildings to cities to the world at large, and at each step, it is generating proprietary maps, from the information networks of Amazon to the street maps of Google. Crawford uses the concept of the atlas to speak to the colonizing impulses that have come to

DOI: 10.4324/9781003225072-5

dominate the AI field, as the science and technologies of applied intelligence and the data sets used to drive it are progressively delimited by a number of industries and agencies that are determining exactly how and why the world is measured. AI, Crawford suggests, is an "extractive industry" (15), resource-dependent, powered by labor, by data and the physical resources needed to advance computational technologies and networks.

Wordbuilding is not simply a matter of informatically driven industrial fabrication; it is also a creative industry, tied to the evolution of video game technologies, and in particular video game engines, including those discussed earlier in this volume: Unity, Unreal and RAGE (the Rockstar Advanced Game Engine utilized in the *Grand Theft Auto* franchise). Advanced game engines are able to handle the coordinated data demands, complex AI arrangements and dynamic environmental effects of large streaming worlds. Open-world games balance the directives of exploration and campaign or mission-based play and the various signposts of persistence that are grown from a base design of the game world at large (Freedman, 2020). The long-running *SimCity* series, first developed by Maxis in 1989 and acquired by Electronic Arts in 1997, popularized open-ended city building, urban simulation and infrastructure management; although the game was envisioned as an opportunity for players to build connected cities, that vision of a seamless collaborative city-building platform with rich multiplayer features never fully materialized. While the 2013 *SimCity* release utilizes the GlassBox engine for dynamic simulation, intelligence has always operated at a fairly low level throughout the franchise, using a basic structure of interlinked game-based decision trees (to build with) and shortest-path algorithms (to steer city residents).

The generative possibilities of video games, game engines and engine-based AI subsystems have laid the foundation for an even broader range of speculative systems—using virtual worlds to make our physical world more discoverable. Virtual worlds are highly coordinated, and they are often developed to realize particular goals. As a story and design tool for virtual production in film and television, game engines (coupled with LED screens and tracked cameras) are also shaping the form and labor of fictional worldbuilding, and advancing the use case for synchronized physical and virtual assets. The competition among those invested in worldbuilding is focused on platforms—which is the most efficient, most user-friendly, most compatible with other software, most open to a wide range of data formats, most adaptable.

Espen Aarseth and Stephan Günzel (2019) have coined the term "ludotopia" to speak to the "dialectical entanglements of games and space" (7), and describe the myriad ways that play transforms the built environment. Play shapes spaces into sites of active practice, into richly grounded experiences; play semanticizes the built environment. But the built environment is defined by systems and organized by infrastructure; it is not an open narrative. Gameplay technologies have been readily adapted to visualize, monitor and test the built environment, actively mapping their playful practices onto everyday life. While it is beyond my purview here to catalogue all of the competing worldbuilding platforms, many of which are still in development, I will lay out several key examples to highlight their most salient features and to tease apart their impact. Worldbuilding systems are built on programming languages and filtered through a variety of development environments and interfaces; those formal frameworks have limited expressive power although they may be well-designed, as they have to hold onto the semantic properties of their core languages and data sets (Felleisen, 1991). At the same time, worldbuilding frameworks are also communication frameworks, and they compete for cultural dominance. My goal is that these case studies, while not exhaustive, can serve as building blocks to comprehend the large-scale environmental impact of artificial intelligence and act as an entry point for understanding these systems and holding them accountable. Every implementation of AI is unique, but as we begin to untangle the material architectures of AI from its contextual environments, tease apart the code, the platform, the data, and the ecosystem, we can start to see its role in shaping the world and how, precisely, it is held under the sway of prevailing social and economic conditions. Here, the passing reference to video game engines is to suggest that their initial industrial context, as development environments for the video game industry, continues to influence their deployment in more expansive realms; and in that context, borne from an industry that largely continues to pursue graphical realism, video game engines are shaping the visual representation of the world at large according to the industrial logic of their game industry origins. The residues of play are found in the ongoing distinctions we can make between production and execution, between development and runtime functions. Commercially viable worldbuilding systems, those that can cast the widest net of possible industrial use cases, are designed to adapt, to prompt discovery and to create playful affordances for their end users.

What is the relationship between worldbuilding and the world? Worldbuilding suggests a studied attention to integrated systems, to understanding how things fit together; and it suggests the principled application of logic. Game worlds, even when they are procedurally generated, emerge from those fundamental assumptions, and AI can only work within the contours of well-defined logic-driven algorithms. Wordbuilding draws from design, programming and engineering, involves creating rules, structures and behaviors, interactive mechanics and modes of exploration; and it also necessitates an understanding of geography, topography, environmental physics, and social practice. The first few chapters of this book consider a number of AI techniques that have been associated with the development of video games, and track the ever-expanding capabilities of AI beyond the circuit of game design and development. I started there for a reason, as I am following a logic that was mapped out rather forcefully by other scholars, including Noah Wardrip-Fruin (2009). If we can understand how and why AI operates the way it does in a fairly obvious visual medium, one of hermetic narratives and largely self-contained data ecosystems, we can use that knowledge to consider how AI is used and why it matters in more pressing and more expansive social contexts and within more sweeping aggregations of data.

Playable Space

In 2018, researchers at Uber AI Labs unveiled Go-Explore and put it to work on two Atari platforming games: *Pitfall* (released in 1982) and *Montezuma's Revenge* (released in 1984). Go-Explore relies on deep reinforcement learning and leverages its family of algorithms to solve what are known in the computer science field as hard-exploration problems—how to explore and fully discover a domain with sparse and delayed rewards that is prone to a high rate of failure (Ecoffet et al., 2021). Go-Explore has since gone on to solve all of the previously unsolved games from the Atari 2600 benchmark of the Arcade Learning Environment (ALE). The ALE is an evaluation platform (more precisely, it is a software framework for interfacing with emulated Atari 2600 game environments) that poses the challenge of building AI agents with general competency across more than 50 Atari 2600 games, and it has led to a number of high-profile success stories in deep learning. Unlike the vast majority of the games in the ALE which are now easily solved by learning agents, *Montezuma's Revenge* had proven resistant to deep reinforcement learning methods, largely because the game's rewards are few and far between, and the game has strict failure conditions; a rather long and complex series of actions need to be completed over an extended period of time before the learning agent receives a reward signal. The game has many rooms, many complex tasks and many unforgiving points of potential death; the agent also has to acquire domain-specific knowledge, understand affordances for (and obstacles to) movement, and distinguish between tools (torches, swords, amulets and keys) and enemies (skulls, snakes and spiders) that are all integral to gameplay. The ALE, introduced in 2012, serves as a bridge between the Atari 2600, first released in 1977, and the development of advanced methodologies for evaluating machine learning. The insights gleaned from ALE research have led the way to building domain-independent AI technologies—AI agents that are capable of broad competence across multiple environments and across varied tasks. That broader competence is grounded in the ability of AI agents to archive previously found states, recall them and then explore from them, and those principal activities form the foundational problem for intelligent agents in reinforcement learning environments; they have to remember where they have been to consider where they are going. This iterative state building process continuously expands the agent's sphere of knowledge. The matter of successful task-oriented progression is complicated by the matter of mapping, which can proceed on purely structural grounds or with additional semantic considerations—options that depend on the environment and whether or not the outcome is time sensitive. The success of Go-Explore is linked to the ability for the program to learn from restorable versions of an environment (stored simulations): "When exploiting this property of restorable environments, Go-Explore thoroughly explores the environment during its 'exploration phase' by continually restoring (and subsequently taking exploratory actions from) one of the states in its archive" (Ecoffet et al., 2021, 582). Go-Explore learns from trial and error until it returns the most favorable trajectory or, more generally speaking, the most favorable outcome. Go-Explore and similar learning algorithms translate environmental exploration into a mathematical problem, although they may do so with different formulas and with varying rates of success.

Why are Uber engineers playing games? These object-oriented applications of artificial intelligence occur in tandem with the integration of AI in the built environment and with its more generalized integration into the flows of the workplace and the marketplace. AI is part of a data-dependent ecosystem and often the central organizing agent, and it is being used to attack a wide range of emergent problems. Still in its infancy, Go-Explore has provided solutions for pre-existing problems, though its value will undoubtedly continue to expand. Currently, it works with established goal-conditioned policies to solve "sequential-decision-making problems" (Ecoffet et al., 2021, 585). The most obvious real-world applications of the Go-Explore algorithm are in robotics, where it can be used to navigate highly dimensional spaces, and in industrial practices that can benefit from machine learning. But the pursuit of better algorithms is also attached to new ways of exploring space, moving beyond intrinsically motivated reinforcement learning that can be understood as playful and curious (in the absence of an explicit reward) to something more goal-directed or purposeful, extrinsically motivated and highly efficient. While intrinsically motivated or playful behaviors are valuable, as they present opportunities to accumulate skills, they are more narrowly or inwardly focused in their serialized pursuits and as a result, they are less robust in contributing to a specific pre-determined goal (Singh et al., 2010). The Go-Explore algorithm, however, broadens the field of exploration, continuously expands its domain knowledge, and becomes more outward as it proceeds, to achieve better results. More immediately for Uber, big data lie at the center of the company's transportation platform, which is grounded on effective exploration. Uber's rider app captures the user experience, as Uber AI works in the background to integrate several common intelligence verticals: computer vision, natural language processing, deep learning and location sensing.

Ludwig, Uber's open-source deep-learning toolkit released in 2019, represents the roadmap for companies that have first leveraged machine learning for their own localized applications: deliver the algorithm to any learning-dependent field. Ludwig sits within the even grander open-source software libraries of a number of machine-learning frameworks, and is built on top of Google's TensorFlow, a machine-learning platform that supports data acquisition, training and predictive analytics. As with Apple's Neural Engine, TensorFlow provides the resources (neural network building and generative adversarial network training) for on-device inference, and can support a range of mobile photo features including image classification and object detection. Ludwig allows users to train deep-learning models without writing a single line of code:

> Ludwig provides a set of model architectures that can be combined together to create an end-to-end model for a given use case. As an analogy, if deep learning libraries provide the building blocks to make your building, Ludwig provides the buildings to make your city, and you can choose among the available buildings or add your own building to the set of available ones.
>
> *(Molino et al., 2019)*

Deep learning has been applied across Uber's holdings, as part of its application build and the ongoing iteration of its movement platform, and as part of its research on self-driving technologies (a division of the parent company that was acquired by San Francisco-based startup Aurora in 2020); and through its data resources and its learning algorithms, Uber is learning more about the physical world to perfect its digital world. Uber's familiar app is simply one vehicle in a densely woven information ecosystem, borne from and dependent on the company's investments in machine learning, and it is used within the company for a number of interrelated tasks including customer support, managing chat functions, improving maps, and building forecast models. Uber's machine-learning algorithms, constellations of complex math and code that improve over time, as they process more data, provide the functional though hidden architectures for the client experience.

Uber stands within the mainstream of technology companies that are connecting the physical and digital worlds, while building a culture that has largely signed off on the collection of personal data to support a robust infrastructure. In March 2020, software developer Niantic, perhaps best known for the augmented reality games *Ingress* (released in 2012) and *Pokémon GO* (released in 2016), acquired spatial mapping startup 6D.ai to further its work in building a dynamic 3D map of the world. 6D.ai has been focused on creating a persistent 3D topographic mesh that can be anchored to its real-world counterpart, respond to occlusion (preserve the rules of lines of sight), and readily sync across multiple users (Fink, 2018). In November 2021, Niantic opened up its Lightship platform (the technology behind its AR games) with a developer kit to expand development and deployment with its AR suite. While real-time mapping is a significant feature of the Augmented Reality Development Kit (ARDK), Lightship also facilitates "understanding" or semantic segmentation, using computer vision to identify the key elements of a real-world environment (land, sky, water and buildings) to properly inform how virtual content should react within a real space (Niantic, 2021).

Augmented reality is only one model of worldbuilding; it presents a useful layered approach to bridging data with the physical world. Other approaches work in an opposing direction and pull data from the physical world to inform a linked but fully realized virtual domain. In December 2020, fashion company Balenciaga partnered with developer Epic Games to debut its Fall 2021 collection through a video game titled *Afterworld: The Age of Tomorrow*; built with the game developer's Unreal Engine, live models were captured on a mobile volumetric video stage. In September 2021, the fashion house extended its partnership with Epic to provide designer skins for several *Fortnite* characters, as part of a sweeping collaboration that also includes limited-edition real-world apparel; the in-game garments were created from 3D scans of Balenciaga's designs and modified for use in the game world. Epic is looking to leverage Unreal, *Fortnite* and its other assets to claim territory in the metaverse, a persistent bridge of the physical and virtual worlds—the next iteration of the Internet, with a powerful data architecture that can further the tenets and experience of immersion while also shaping online commerce. Matthew Ball sums up the primary attributes of the metaverse with a rather precise, though perhaps preliminary, definition:

> The Metaverse is a massively scaled and interoperable network of real-time rendered 3D virtual worlds which can be experienced synchronously and persistently by an effectively unlimited number of users with an individual sense of presence, and with continuity of data, such as identity, history, entitlements, objects, communications, and payments.
>
> *(Ball, 2021)*

Game engines and their attendant data infrastructures are the building blocks of the metaverse, and developers such as Epic and Unity are using their proprietary engines, assets and data holdings to claim territory, buying up competing or complementary technology companies to secure hermetic worldbuilding systems; in 2019, Epic acquired Quixel and its photogrammetry asset library (a world atlas of high-resolution 3D scans of environmental specimens—surfaces, materials, objects and vegetation) and in 2021, Unity announced its plan to acquire Weta Digital and the company's proprietary visual effects suite, stating its intent to "deliver tools to unlock the full potential of the metaverse" (Whitten, 2021). Unity's acquisition of Weta Digital includes a robust asset library of "urban and natural environments, flora and fauna, humans, man-made objects, materials, textures, and more" (Whitten, 2021). In this closed circuit of relations, human assets are being fabricated from the same algorithmic architecture as their virtual worlds. They are being constructed from and operating within the same governing data structures. The applications of real-time 3D visualization are grounded in the industrial legacies and logics of video game design, and that particular programming impulse is influencing the broad swaths of visual culture that have been swept up by creative worldbuilding.

Video game engines are powering our visual futures, and as data and visuality continue to converge, Epic, Unity and other software developers are rapidly iterating their products to tackle new markets; they all have plans to colonize the metaverse with their proprietary data holdings, design tools and aesthetic vocabu-laries. Beyond their transformation of distinct media verticals, Epic and Unity have a shared pursuit: build the metaverse and populate it with human assets. That intent is grounded in marketing materials that posit ideal use cases, and in software documentation that, for each company, lays out a series of foundational skills and concepts and illustrates these through a number of use-based video training modules and proof-of-concept demonstrations that are informed by the forms and functions of video gaming. I use the concept of colonization here to call out the organizing power of dynamic real-time 3D visualization, the steady acquisition of complementary data holdings to form hermetic systems, and the linguistic conventions (the programming, interfaces and representations) that are grounded in the hyper-realistic pursuits of video game development; extending Crawford's model, we can label these operations as both regulative and extractive in their technical work and in the work they perform on representation. We need to attend to the origin stories of Epic, Unity and a number of other engine-based developers if we are to understand what is driving how we represent the world and identify the traces of a particular industrial process that grounds its represen-tations in the aesthetics and applications of video game development. We see the influence of Epic and Unity in their engines and in the ways both companies envision and embody their use. The colonizing influence of platform developers lies in their ability to get us to choose one development language, one build envi-ronment over another, as they create a market for their industrial products and their systems (and assets) of cultural representation. The sociotechnical imaginar-ies of simulation guide how we articulate and develop the physical world, and the dominant simulation technologies are pulling together distinct domains of expert practice (media, technology, engineering, medicine, law) that have heretofore operated with their own imaginaries about how to organize and develop the world. Informatic dominance is tied to political, economic and cultural power and secured through a number of licensing agreements and competitive acquisi-tions. Simulation, as with computation and machine intelligence, operates across many distinct industrial vectors. Broadly speaking, we understand what it means to simulate an object, environment or experience, but we can task a computer with building a simulation of an almost infinite array of objects, environments and experiences. The power of programming lies in its transmutability or the ease with which it can reproduce such diversity. As we attend to how simulation is applied, we should be mindful what, beyond objective experience, is being expressed, how as Noah Wardrip-Fruin (2009) suggests, the digital artifact may be an operationalized model of its subject, expressing a particular design intent or any number of client-centered priorities. In the company's press material for the Balenciaga project, Epic Games suggests that *Afterworld* represents an opportunity to use digital human avatars to "reset identity" (Epic Games, 2021). Players can pick from a range of characters as they navigate the game's futuristic city and peruse the French label's Fall 2021 collection. At CES 2022 (an annual consumer electronics show), Hyundai Motor Corporation and Unity announced a partnership to design and build a Meta-Factory, a digital twin of an actual factory supported by a metaverse platform: "The introduction of a Meta-Factory will allow Hyundai to test-run a factory virtually in order to calculate the optimized plant operation, and enable plant managers to solve problems without having to physically visit the plant" (Hyundai, 2022). The power of programming is not simply tied to progressive transmutations and "identity resets." It is also tied to very real material conditions that are shaping the structures of labor; as a digital artifact, an intelligent twin of factory operations, the Meta-Factory rather openly operationalizes its human subjects within a model of ever-increasing workplace efficiency.

Modeling With Intelligence

Digital twins, virtual representations of real-world entities and processes, are data-intensive structures; simulating real-word structures and systems requires an enormous amount of compliant data. Real-time 3D technology and AI are being leveraged among competing (and occasionally collaborative) stakeholders, including Epic, Unity, IBM, Oracle and Microsoft, and the key to market share is about both proprietary technologies and data. Each of these companies has the massive data holdings that are necessary to produce accurate real-time 3D visualizations—dynamic renderings that have become central to the place-based practices of architecture, engineering and construction. Unity, for example, allows its customers to pull data from sensor-based systems into its cross-platform game engine, to produce interoperable location data, CAD data, computer vision data and natural language processing data (D'Anastasio, 2022). Digital twins use real-time and historical data to represent and simulate past, present and predictive future system states, and are built on data, tailored to meet particular use cases or scenarios, and guided by localized domain knowledge. Digital twins are built from a three-layer architecture: the physical hardware layer supports the operational functions of computation, communication, and data storage, the virtual layer of the twin itself captures and reflects the real-time data and analytics of the physical world, and the interaction or command layer serves as a bridge that connects end users to both the physical world and its twin.

Together, the expansive computational development of digital twins and the technical practices of engineering and construction are impacting the social practices of architected space and a number of distinct institutional environments—our homes, schools, hospitals, transit hubs, workplaces and cities—that are dependent on common components of infrastructure (including energy and data). Beyond the built environment, digital twins have been critical to a number of other industrial vectors such as aviation, aerospace and defense, where digital technologies have been leveraged to collapse the cycle of designing, prototyping, testing and building by enhancing digital model-based systems with data analytics and simulation. Aerial photography and mapping, complemented by robust 3D visualization and graphics tools, are being used to visualize architecture, infrastructure and the built environment. Technology is embedded in the design process, shaping its workflow and outcomes. Digital twins have become a full-fledged industry, focused on building digital representations of physical assets that can retain a live connection to the world they represent; when used for planning and assessment, digital twins provide a framework for creating a digital truth about the physical environment and are used to draw out the cause-and-effect relationships of something that is happening in the built environment. Digital twins are not simply a building information model or a tidy visualization of a database, but something more diagnostic and disruptive. The digital twin simulates and learns from its real-world counterpart(s); it is process-oriented and responsive to real-time data. Digital twins are not composed of a single entity and not driven by a singular motivation; they are built from multiple components and have multiple data layers, most of which draw from an IoT. While we can decipher how digital twins are resourced and constructed, within this shared data space what may be less obvious are the range of interests (public, private, civic and corporate) at work—interests that may privilege different components and walk away with divergent readings. To understand the value of digital twins, we need to consider their technical processes, their data points and the various motivations that orient their modeling. We need to consider how the data that drive digital twins are collected, stored and activated, and how it is developed into particular 3D models or representations.

In 2019, Epic acquired Twinmotion from developer Abvent, adding architectural visualization to the Unreal Engine suite. The software allows designers to work with BIM (building information modeling) and CAD (computer-aided design) data in Unreal alongside the engine's other prebuilt assets (including people and vegetation) and environmental effects. Artificial intelligence provides a resource for automating parts of the visualization process, to identify, segment and classify objects within a 3D mesh and add critical attributes such as building heights, volumes and materials. AI also adds a layer of intelligence to the foundational geometry of semantic 3D models by organizing an actionable data layer that can be used for forecasting and analytics; most digital twins tie their semantic 3D models to 3D mesh models, connecting simple geometries to a continuous photorealistic mesh while allowing the underlying data sets to flow between these layers.

While we might suggest that the migration of game-based technologies into non-game industries is evidence of the increasing gamification of culture, more important than the "use of game design elements in non-game contexts" (Deterding et al., 2011, 9) is the use of game software in non-game contexts—a use that furthers the "platformization" of culture. Go-Explore and Unreal present two rule-based approaches for exploring space with an action-oriented or goal-based perspective; and while one is focused on discovering and the other is focused on building, they are both in sync with the behaviors of a platform society. Digital twins have been made possible by ubiquitous computing; they are a by-product of the rampant proliferation of computing in the physical world, a concept introduced by Mark Weiser in the 1990s. Yet Weiser's vision of ubiquitous computing (1991) speaks to the operations of users in environments that are augmented with computational resources and not the long data trails from the IoT—a grander vantage point that aggregates data from devices and users to develop an informatically rich simulation. Ubiquitous computing has led beyond networked mobility to the full-fledged technologization of urban space and heightened possibilities for semantic modeling; the technologies of municipal planning, focused on improving administrative workflows and guiding the utilization of city resources, have been developed into a series of interlocked systems and data sets that, when properly aligned, can drive regional decision-making. Cities have expanded their infrastructures and their capacities to generate data; they have become integrated municipal and industrial platforms that operate across every economic sector and influence every sphere of life.

As their use has become more commonplace, digital twins have been put to work in complex ecosystems to tackle real-world problems. The New South Wales (NSW) Spatial Digital Twin, part of a government strategy for building a unified spatial data framework, is a digital tool for guiding infrastructure planning and management, land use, and other collaborative data work, and has been designed to create attachment points between government, industry and area residents. The platform allows the government to visualize historical considerations and model future scenarios, and includes a comprehensive inventory of existing above- and below-ground infrastructure. The government of New South Wales partnered with Aerometrex, a private firm headquartered in Australia, to capture and process the necessary 3D data to underpin the NSW Spatial Digital Twin, gathering information from three-dimensional aerial imagery and LiDAR (light detection and ranging) surveys and integrating it with the state's land parcel registry. The platform also integrates digital engineering assets (surveys, plans, design reviews, contracts and object models), BIM resources and live API feeds for public transport, air quality and energy production; the platform spans multiple municipal sectors and ties the built environment and its infrastructure to the natural environment (New South Wales Government Services, 2022).

The Wellington Digital Twin, a digital twin of Wellington, New Zealand, is a large-scale visualization of the city informed by an array of integrated real-time IoT, geospatial and terrain data as well as detailed photogrammetry models that define the footprint of each building. Built on Unreal by New Zealand-based studio Buildmedia, the platform can ingest multiple data file formats to bring the twin to life and visualize the complex interrelationships of city-wide activities and systems; the twin can also help developers consider the impact of proposed infrastructure and building projects on the city at large, visualize and communicate these proposals to stakeholders and illustrate how these might integrate with the existing built environment.

FIGURE 5.1   Wellington Digital Twin developed by Buildmedia on the Unreal Engine, highlighting the projected role of mass transit.

Source: https://buildmedia.com/work/wellington-digital-twin

Copyright 2022. Buildmedia

FIGURE 5.2   Wellington Digital Twin developed by Buildmedia on the Unreal Engine, highlighting cyclist sensor data.

Source: https://buildmedia.com/work/wellington-digital-twin

Copyright 2022. Buildmedia

Beijing-based 51World has built digital twins of Shanghai and Singapore using the Unreal Engine; and at a more micro-level, the company has used the engine to create a digital twin of the Wuyi Square subway station in Changsha, Hunan, China, as part of an effort to study and improve passenger flow. The latter project pulls live data from sensors in the station that can track the movement of people, transit machines and trains, and the twin can use AI to simulate crowd behaviors in order to stress test a variety of environmental scenarios.

London-based architecture firm Foster + Partners partnered with Boston Dynamics to use Spot, the robot dog, to assist with construction at Battersea Power Station in London and to test its utility in a number of dynamic environments. At Battersea, the firm deployed Spot to explore, collect and feed data back to the company to update the digital construction twin, a four-dimensional representation (of space plus time) of the project that could convey how the site changed over time; Spot was used to repeatedly capture a fixed set of on-site laser scans while autonomously navigating a pre-recorded path through the construction site.

In the Battersea application, as well as the projects in Australia, New Zealand and China, integrated information on buildings and infrastructure can be paired with artificial intelligence and machine learning to guide ongoing planning, resource management and development. The research team at Foster + Partners notes: "By combining temporal and spatial information with data from sensors that read environmental conditions and occupancy we can construct intricate models of how people, furnishings and environmental conditions interact to form a comprehensive whole" (Tsigkari et al., 2021). The AI-powered digital twin provides both a macro- and a micro-view on our place in the built environment, and not only visualizes but also activates the interdependencies between distinct data sets and can project their futures. Yet digital twins remain bound to a number of external systems that shape how they produce meaning; they need to be realized in a platform. While many of the aforementioned city builds were executed with Unreal, the Battersea project was dependent on the Avvir reality analysis platform, a system focused on the comparative analysis of sequential BIM models (Tsigkari et al., 2021) that allowed the team to track the progress of construction. Digital twins are built from a series of proprietary constructs; they exist in platforms and run on hardware that has to negotiate between distinct data formats. Digital twins rely on an enormous amount of data about the physical world, pulled from localized sensors and global satellites, and combine data from multiple sources, including computer aided design data and IoT data from networked devices, to build complex multisystem environmental models; they converge what were once distinct industry workflows. Within these models, AI is not an isolated immaterial practice; as an assemblage of collected and activated data, it is dependent on physical resources and objects. To know the world, the digital twin has to connect with the world, and it does so through a number of embedded machine-readable vantage points.

On a more localized level, platform developer Matterport has found its niche in 3D capture as an integrated service provider for the real-estate market, using its proprietary Cortex AI to produce dynamic digital twins. Matterport's holdings include a patented deep-learning neural network; this walled garden of aggregated spatial data can be enhanced with the company's computer vision tools and cleaned up with its proprietary image processing tools. Matterport, which has positioned itself as a fully integrated cloud service provider, represents one of many flashpoints in the development of closed AI-as-service systems—systems that seem counterintuitive to the operational logics of informed and inclusive intelligence and seem to be at odds with the open data systems being developed by the government bodies I have mentioned here. Matterport's focus has been on digitizing the built world, a business model grounded on reality capture that has developed into a closed self-monitoring system, designed to confirm the world as we know it.

Playful Playability

So how precisely do digital twins fit within the framework of playable media? Are digital twins responsive, discoverable media systems? Do they invite playful playability? Counter to the impulse most notable in Matterport's approach (how and why it trades in data), playable media are performative and relational rather than transactional; they exist at the boundaries of systems, and they play with ideology. Playable media are generative and responsive and use contextual awareness (in the technical sense) to promote situational awareness (in the practiced, social sense). Artificial intelligence is at times used rather unidirectionally and without deliberative agency—a steadfast march through data toward human-level intelligence. Moreover, AI systems have been designed to match the way we, as humans, organize the world, even as they do so more efficiently. Are digital twins playful and playable? As with their AI subsystems, the answer is grounded in both their form and function. How are particular digital twins constructed and what are they tasked with doing? Who is playing them?

This manuscript is grounded in the idea that playable media can produce valuable meaning through their mechanics, and we can advance that concept beyond proprietary systems and serialized intellectual properties by considering a number of AI applications that find value in responsive worldbuilding—value we can see in the governmental applications developed in Australia and New Zealand that have connected data visualization to an accessible public interface tool. I use the concept of playable media throughout this manuscript to suggest a series of situated practices and reciprocal engagements that coalesce into a system, that allow physical design to respond to human relationships and that use data to act; this can hold true even if a system has limited rights of access and is conceived for a restricted user group. As I suggested at the outset of this project, playable media make the world more dynamic and discoverable. We do not need to break our existing intelligence systems; we simply need to be more thoughtful about how we stitch them together as we build more open media architectures. The aggregated data projects undertaken by government agencies in Australia and New Zealand have led to digital twins, as part of parallel efforts to be more rather than less communicative. One notable feature of the NSW Spatial Digital Twin is its accessible network dashboard, an interface that allows users to openly explore map data, parse through a series of visual layers, and add stories—annotations to individual map scenes that can build narrative context for the network of spatial data.

As an extension of AI within platforms, as self-contained knowledge-building systems within intentionally designed ecosystems or as living references to physical structures, digital twins present a new frontier for AI, bridging the physical and the virtual world and seamlessly transmitting data into a virtual entity that can co-exist with its physical counterpart; a digital twin is a digital duplicate of a physical object or system, and can be used for system design and operation, to observe and monitor system performance, represent historical states, explore or predict future states, and train. Digital twins are not static entities, but are virtual representations of the real world that can incorporate physical objects, as well as information about processes, relationships and behaviors (Andrews, 2021). This, in effect, is the concept of twinning rather than replacing the original living or non-living physical entity. Digital twins have had their most obvious impact on architectural visualization, to render buildings intelligible, but they are also influencing wider scale smart city development by pulling together geospatial technology and GIS data, building information modeling and tools for interactive 3D visualization. Digital twins can simulate the imprint of entire cities, the interdependencies of the natural and built environment, as well as the stress points between mechanical processes and human activities. Digital twins are generated from a set of interlocked technologies and have been enabled by several interlocked technical levers: artificial intelligence, the IoT and cloud services. Though digital twins have developed the furthest in manufacturing, similar investments have expanded within the aerospace and defense industries. Northrop Grumman, as one lead example, has significant investments in both digital twin and AI technologies.

David Gelernter captures the idea of digital twin technology in *Mirror Worlds* (1991) and describes persistent data-rich human-scale software models that reflect the world, its objects and its systems. Digital twins had their first industrial applications in aerospace. NASA's interest in digital twins was motivated by its requirement to operate, maintain and repair remote physical systems. During the 1970 Apollo 13 mission, NASA retained a mirrored system on earth that allowed engineers and astronauts to determine how they could safely bring the crew home, continuously streaming data to modify its simulations in order to reflect the evolving conditions of the crippled spacecraft (Ferguson, 2020). The mirrored systems deployed by NASA can be considered precursors to modern digital twins; while their technical foundations are not identical to current methods for

3D visualization, they performed the same functions—managing and modeling complex systems, monitoring their operational health, and developing predictive scenarios based on data driven analytics.

While the work at NASA was grounded in data-rich simulations, the conceptual model of the digital twin was formalized by Michael Grieves in 2002 and grounded in the field of product lifecycle management (Grieves and Vickers, 2017). Grieves proposed physical and virtual systems that would share data over time; the mirrored spaces model, also referred to as the information modeling model, was not a static representation of these systems, but one that connected them over an entire manufacturing lifecycle (Grieves and Vickers, 2017). The arrangement ties together a digital information construct, realized as a virtual system, to its physical corollary over its full lifecycle. The value of the concept, while initially applied in the context of manufacturing, has expanded into other industrial fields. The nomenclature was formalized in a 2010 NASA technology area roadmap report for "Materials, Structures, Mechanical Systems and Manufacturing," and advanced through the concept of a Virtual Digital Fleet Leader (VDFL):

> The VDFL paradigm integrates (cross cutting) multiple technology capabilities, into a multi-physics, multi-scale simulation of the as-built vehicle or system. The VDFL incorporates high-fidelity modeling and simulation and situational awareness into a real-time-mission-life virtual construct of the flying vehicle or system. The VDFL continuously forecasts the health of the vehicle or system, the remaining useful life and the probability of mission success.
>
> *(Piascik et al., 2010, 2)*

Most modern digital twins involve synchronized physical and virtual structures that are dynamically connected through a continuous stream of data; the ongoing flow of information is used to update the digital model, its structures and behaviors in real time and maintains a converged state between the physical systems and assets and their simulations. A digital twin is not a static entity, but requires persistent twinning; as Ben Hicks (2019) notes, "It is this concept of continuous or periodic 'twinning' of the digital to the physical in order to mirror condition that separates virtual prototyping and model-based engineering from a digital twin." As digital twin technologies have evolved, their horizons have become grander; no longer bound to object-oriented systems, digital twins are swallowing cities whole, ingesting larger data sets, managing both quotidian activities and lifecycle events, and monitoring the relationships between humans, machines, built structures and the natural world.

Simulated Cities

The push of artificial intelligence and machine learning, coupled with developments in computational hardware and cloud services, has expanded the applications and values of digital twins, which have become central to such human-centered processes as decision-making, risk assessment and collaboration and such physical processes as the design and development of the built environment, and the production of smart cities saturated with and mobilized by data. Digital twins can be used to visualize unbuilt infrastructure or properties, and coupled with smart city technologies to inform planned improvements to existing structures, but these visualizations remain tied to and conformed by platforms.

Esri's (Environmental Systems Research Institute) ArcGIS mapping and analytics platform—a software tool that translates data into location-based insights—has been used as a foundation for a number of digital twin builds including the Amsterdam Airport Schiphol, the Hartsfield-Jackson Atlanta International Airport and the Long Island Rail Road, all focused on modernizing infrastructure and deploying digital twins to address the needs of multiple stakeholders: designers, contractors, engineers, regulators, occupants and passengers. These efforts are, of course, tied to a central business proposition. IBM, which supports the digitization efforts in Amsterdam, reminds its clients that digital twins can be used to "run simulations comparing as-is and what-if processes to calculate ROI. Then refine models and disqualify unproductive changes" (IBM, 2021). Here, artificial intelligence plays a role to recognize patterns and capture them as actionable data as part of the larger enterprise of process mining (using data mining algorithms to understand and improve employee workflows).

A number of companies have turned to Unity and the Unity engine to build digital twins and have collected synthetic environmental or functional data from these simulations to advance their real-world doubles. Unity has partnered with several international airports, including Hong Kong and Vancouver, to build environmental simulations that can track passenger traffic flow in real time, and has extended its engagement with the Changi Airport in Singapore to gamify the experience. Year 2019 marked the opening of the Changi Experience Studio, an interactive attraction that invites visitors to go behind the scenes of airport operations; a series of four Unity-based "efficiency games" simulate several of the processes and challenges of airport staff, including baggage handling. The digital twin of Hong Kong International Airport (HKIA) integrates the building information models of HKIA with various operations data from the airport's asset management system and localized geographic information systems data; the model is fed with real-time data from IoT devices deployed throughout the airport. The twin is coupled to machine-learning and analytics tools and can run simulations of terminal performance under various conditions (such as changes in passenger traffic volume) to optimize existing operations and streamline the review of future projects (Wordsworth, 2019). Each of these digital twins is grounded in a defined architectural footprint and can integrate information about the building envelope with its infrastructure, assets and functional systems; and many of these twins are part of a larger AI strategy. In addition to its digital twin, the implementations of data-driven intelligence at HKIA include advanced biometrics, video analytics and robotics, all targeted at security, management and airport operations.

In 2015, Dassault Systèmes announced a collaboration with the National Research Foundation Singapore to build Virtual Singapore, a citywide digital twin. Known as 3DEXPERIENCity and powered by Dassault's proprietary 3DEXPERIENCE platform, the project provides Singapore with a unified 3D model to assess and solve a range of urban challenges, as it pulls together the 3D efforts of various public agencies (and data collected from devices worn by students) into a unified platform. Designed to amplify the work of city planners through an integrated reference model, 3DEXPERIENCity also functions as an authoring environment to run if/when simulations and spot trends that can inform ongoing urban planning and development. Virtual Singapore and related city platforms that can capture and integrate environmental data and use that data to run new programs, introduce new parameters and test new scenarios seem to be moving their users toward a unified truth and a proven model for the future that is supported by data and affirmed through a series of visual reference points that readily validate the collective desire for a holistic view of the built environment. That truth can be traced back to the hard data embedded in semantic 3D city models and building information resources that capture information from the built environment (buildings, streets and bridges) and other physical entities (such as trees and other vegetation); and the necessary balance between competing interests (in building terms, these often yield to the common pursuit of net-zero construction and the creation of sustainable infrastructure and cities) can be found through algorithmic manipulation.

While there have been several field-specific approaches to urban informatics, attached to distinct enterprise applications in construction, gaming, urban simulation and geomatics (Kolbe and Donaubauer, 2021), these approaches seem to be converging within a few select platforms that can present advanced 3D geometry and readily decomposed 3D models, and accommodate a detailed information model with clearly classified spatial and mechanical data; visualization and computation need to peacefully coexist in the digital twin. Thomas Kolbe and Andreas Donaubauer (2021) note the particular organizational challenge of building an urban digital twin: "In contrast to industry, where all the information about a specific product is bundled by the manufacturer, the information about real-world objects of cities like buildings, streets, bridges, and so on is distributed across several organizations and stakeholders" (630). The central challenge for city planners is collating what has commonly been distributed knowledge about the city, and the central challenge for platform developers is aggregating, transcribing and synthesizing disconnected volumes of heterogeneous data to produce a unified view of a city's present and an operational unit model that can chart its future.

In the closing chapter of *The Death and Life of Great American Cities*, influential urbanist Jane Jacobs quotes Warren Weaver in a 1958 Rockefeller Foundation Annual Report, noting Weaver's proposition that scientific progress has unfolded over three distinct phases: addressing problems of simplicity, then problems of disorganized complexity, and finally problems of organized complexity (Weaver, 1961). Progress beyond the first stage required the introduction of theories of probability and depended on statistical methods to understand large systems that have many variables. Problems of organized complexity, of organic wholes, of biology, required new ways of seeing and new forms of analysis.

In an age of unprecedented computing power, a host of problems that were previously only understood through disorganized complexity are now becoming discoverable. What does this have to do with cities? Jacobs puts forth that cities are essentially problems in organized complexity. Cities are situational, have many interdependent variables and are replete with relational nuances: "Cities, again like the life sciences, do not exhibit one problem in organized complexity, which if understood explains all." (Jacobs, 1961, 433). Cities are intricately interconnected, but those relationships can be untangled, examined and understood (Jacobs, 1961). Jacobs pushes further to consider the dynamic nature of cities and suggests that to understand cities, we need to think about processes (Jacobs, 1961), of the relational values of city assets (e.g., housing units) and how these assets can be deployed as catalysts for change.

Cities are not simply sum totals of their infrastructure or aggregates of data collected from an IoT. They are distributed architectures of media and information and defined by the flow of ideas. While technology has worked its way into our bodies, our objects and our environments, the effect has not (as of yet) been to create a parallel, simulated world but rather an embedded world (De Monchaux, 2021). To develop models of increased efficiency and predictability, to leverage digital twins to future-proof and future-cast, is a new variant of what Jacobs refers to as social taxidermy:

> To approach a city, or even a city neighborhood, as if it were a larger architectural problem, capable of being given order by converting it into a disciplined work of art, is to make the mistake of attempting to substitute art for life.
>
> *(Jacobs, 1961, 373)*

Machine intelligence has the potential to over-render organic possibility.

Researchers at the Department of Energy's (DOE) Oak Ridge National Laboratory (ORNL) have developed a modeling program that provides energy details for every one of the 129 million buildings across the United States using publicly available data. When the project began in 2015, there was no single resource to accurately visualize and quantify energy details for each building. The modeling software suite, known as Automatic Building Energy Modeling, or AutoBEM, was developed to achieve ORNL's five-year Model America vision: to generate digital twins for every building across the country and attach these to building energy models. The grander goal has been to digitalize the underlying infrastructure of the built environment and realize it as a series of visible processes that can be used to fight climate change. Digital twins are repeatedly constructed as purposeful and intelligent models of change, dynamic enhancements to otherwise static building information systems and necessary platforms for operationalizing smart cities that can lead to more intentional and less harmful decision-making.

Digital Twins as Theory and Praxis

As buildings, cities and worlds become systematically mapped or twinned in ways that account for structures, systems and inhabitants, and in ways that are also dynamic, unfolding and aware of their influence on other bodies, mental and physical space seem to be converging, as mental maps are outperformed by the nature of visible, if virtual evidence. Mental space becomes less useful as a theoretical practice if algorithms can disprove its value. In *The Production of Space*, Henri Lefebvre (1991) proposes a "unitary theory" of space that can account for the physical, mental and social conceptions of space—an accounting of the physical nature of things, the operations of imagination and abstraction, and the experience of habitation or practice, as space is not simply measured or perceived, but acted out and performed. Digital twins form a unitary praxis that ties together representation, performance and perception, and they accomplish this through data; they allow us to see space as a process, and their exactness and dynamism (using data from multiple sources, most digital twins utilize machine learning, artificial intelligence and advanced modeling techniques to continuously learn and update themselves to represent the current state of an object or process) make them something grander than reference models. Digital twins utilize AI agents to run future-casting simulations and identify patterns in existing systems that can be used to prescribe actions and mitigate challenges in the physical world.

If video games entangle space and play and indeed give meaning to space, as Michel de Certeau (1984) suggests as a "practiced place" (117) of complex intersections, and if video games and digital twins share a common software architecture, are digital twins imbued with the same playful properties as video games? Do they qualify as playable media? We can certainly acknowledge that digital twins are models of spatial praxis; the goal of twinning is not stasis but lifecycle, mapping the status of an object or system over time. De Certeau suggests, "the street geometrically defined by urban planning is transformed into a space by walkers" (1984, 117); it is given a temporal dimension, and produces what digital twin developers refer to as a four-dimensional representation of space plus time. But the formation of a digital twin is less an act of reading (while walking) and more a matter of fluidly automated transcription. Digital twins seem to represent a frontier for colonizing and hegemonizing space (or place-based practice), taking ownership of its data, aligning it with the industrial logic of efficiency, and transcribing it through the functional and normative languages of 3D visualization; human traffic patterns are simply one set of aggregated (mobile) vectors, woven into the IoT to inform the simulation. Digital twins are forged by convergent information models: composites of physical structures (realized through advanced computer graphics), existing operational data and processes, and GIS (geographic information system) and BIM (building information modeling) data.

Responsive worldbuilding requires a close attention to situated practices and reciprocal engagements that allow physical design to respond to human relationships (Freedman, 2020). In *A Pattern Language*, architect Christopher Alexander (1977) argues, in part, against the narrowing frame of computer-assisted architectural visualization. When more openly scripted, coding can make design systems more nuanced and complex; artificial intelligence can recognize patterns and reveal formerly unseen possibilities that, in practiced application, can shape physical structures. Digital twins are built with software, which itself is a place where code has coalesced into action as a program, as a field of stable relations that have certain executable functions. With that fixity in mind, with the work of programmers moved from concept to execution, we can start to perceive the values of code and the material outcomes of algorithms. Adrian Mackenzie (2006) suggests that software serves three explicit social functions: as a determinant of agency, as the foundation of materiality, and as an organizing force for sociality and collectivity. The software architectures of digital twins "rematerialize" (Mackenzie, 2006, 173) existing physical structures and affect how we see, experience and imagine them; within these virtual structures, it becomes quite possible to radically shift sociality and transform relationships without any immediate material consequence. Digital twins are developed to assess and mitigate risk, but they are also promoting a new integrated systems literacy that is becoming an important element in finding our way around the physical world. The machinery of connected data is a new form of infrastructure.

While digital twins are commonly associated with richly developed large-scale exterior analyses, equal attention is being paid to visualizing more intimate spaces through machine learning. Unity Technologies has introduced a number of computer vision solutions for home interior applications, encapsulated in a suite of synthetic data generation tools and services and 3D content libraries (Unity Computer Vision Datasets) for generating photorealistic environments. These data sets are not simply used to build static environmental models; they are also used to train, to assist in refining computer vision and the performance of home automation systems (robotic vacuum cleaners, home monitoring systems), and they are built from multiple types of ground-truth that have come to be associated with domestic space, domestic comfort and domestic security (Kamalzadeh, 2021).

Working within the same platform of Unity engine-based development, these environments can be populated with the assistance of machine intelligence. PeopleSansPeople is Unity's open-source human-centric synthetic data generator, released to the public in early 2022; it contains simulation-ready 3D human assets coupled to a parameterized lighting and camera system that can be inserted into a range of 3D visualization projects. PeopleSansPeople can modulate the appearances of its virtual people to create more customizable data sets, and can train its models to adapt to a specified environment and generalize to that domain (Ebadi et al., 2021).

If software systems can transform sociality and shape our perceptions and interactions, we can expand on the common distributive power (over agency) between two technological lineages: the engine-based transformations of the city in the post-apocalyptic fiction of *The Last of Us*, which draw from the practices and tenets of the video game industry and of storytelling writ large, and the engine-based transformations of the city enacted by digital twins, which tether those tenets and practices to other industrial vectors. The tension I hope to have evoked throughout a manuscript on playable media is between two opposing readings of artificial intelligence: machine-learning algorithms that can recognize and exploit the unconscious patterns of human behavior, and the productive and unpredictable dimensions of play that exceed the limits of those algorithms. The latter prospect seems more likely if we concede the importance of algorithmic literacy, understand how to play within systems, and mindfully contextualize our encounters with AI. Perhaps we place too much faith in computation, because we have developed so many dependencies on its many material forms. Ed Finn (2017) proposes:

> The lesson is that it's much harder to question a set of ideas when they are assembled into an interconnected structure. A seemingly complete and consistent expression of a system of knowledge offers no seams, no points of access that suggest there might be an outside or alternative to the structure.
>
> *(10)*

Algorithmic models of culture are quite seductive; this is why I have perhaps overused the term "ecosystem" to capture both the entrepreneurial mindset that produces hermetic data-driven systems, where algorithms have a material imprint, and the comfort of playing within these systems, where well-designed interfaces clean up our data and create harmony.

As they commonly draw from the engine-based architectures that support the development of video games and their runtime functions, we can consider digital twins as play-adjacent technologies; yet digital twins are borne from a distinct business case, designed to assess and manage risk. I have purposefully sidestepped discussing playable cities—those playful yet important citizen engagements with the built environment, often realized through elements of smart city infrastructure (such as sensor and actuator technologies). Smart cities are digitally enhanced physical worlds, and there is considerable speculation that game environments and digitally enhanced real worlds will continue to converge over time (Nijholt, 2017). But it has been my intent here to illustrate how simulation can recontour cities, pulling in real-time sensory data to push out a new and superseding planning and development model. Digital twins are built from the work already performed on cities to make them intelligent. They extract that intelligence and make it performative, displaying it in a dynamic real-time 3D render. Pervasive games or urban games that change the city into a playful "gameful" city take IoT data in a decidedly different, perhaps more democratic direction. Yet many of the citywide experiments with digital twins share the same goal of making the city more open, more knowable; they remain committed to open-source data and to operating in the public interest. These may not be playful interventions with city data, or playful hacks of smart city technologies, but they are closely related, dependent on much of the same data and infrastructure.

It is easy to point to any number of products and services that encapsulate the fundamentals of proprietary AI research and have been central in shaping what we know about the world. The natural language interfaces of Amazon Alexa, Apple Siri and Google Assistant have cultivated radically integrated fiduciary subjects, but the world is infinitely more complex and its resources are unevenly distributed. Digital twins are modeled and visualized as human-centric media forms; they are, after all, designed to engage key stakeholders. Spaces are twinned to create a single source of truth; that imperative is both functional and ideological, and it drives the specific implementation of their associated software and hardware. Destination Earth, a project even more ambitious than Model America, is an initiative of the European Commission that plans to create digital twins to model, monitor and simulate the Earth's various systems and assess the ongoing impact of human activities on the natural world. The project is intended to support a unified effort to combat a number of complex environmental challenges (European Commission, 2021), and the end deliverable is a full digital replica of the Earth, modeled through a number of converged digital twins (drawn from different domains of Earth science), to be produced by 2030. It remains to be seen exactly what can be twinned and what proves too complex for this proposed simulation, but in general terms the functional goal of this AI-powered digital twin is to make the world more intelligible.

Digital twins are textual instruments; they are designed with intentions, yet they are also playable and affective, as they continuously negotiate between their textual and graphical logics. They are far from static systems, and they are more complexly interactive than video games. While I have suggested that players can study and learn the pattern languages produced by artificial intelligence in games, our reading shifts within the digital twin. We may be aware of its operational intelligence, but we are more focused on its outputs, accepting with confidence that the data layer and the visual layer have been properly linked. I am not suggesting that the user is deeply integrated in the meaning-making process; there is no explicit freedom in the digital twin for the user to physically change the discourse (Aarseth, 1997), although the twin is still a model of interactivity. Various scenarios can be run through the simulation, but these are always driven by algorithmic determinacy and responsive to data rather than personal agency. In our consideration and engagements with playable media we encounter a terminal point that seems antithetical to the pleasures we normally associate with play. But perhaps we have encountered a limit that can be defined as an entry point to serious play through playable media. Playable media invite us to play within discursive, experiential systems and limits, but as we do so, they lead us elsewhere. They produce new knowledge, however trivial or profound. Digital twins have a dynamic nature, shared by simulations more broadly and games more narrowly, and generate physical and dialectical entanglements (Aarseth and Günzel, 2019) between present and future material states. Playable media are transformative spaces activated by computational intelligence; in the case of digital twins, the formal transformations we evidence within them are intended to inform our actions beyond them. We can extend our argument to suggest that artificial intelligence in playable media is persistently performative and inherently transformative; it reads us a series of inputs in an informatic system and situates us as part of its broader organizational schema. Playable media invite us to become part of their computational structures and are, in that way, part of an ongoing push toward computational convergence; in their form, though perhaps not their immediate function, playable media serve as a primer for imprinting ourselves in the persistent worlds of the emerging metaverse.

BIBLIOGRAPHY

Aaltola, Mika. (2012). *Understanding the Politics of Pandemic Scares: An Introduction to Global Politosomatics*. New York: Routledge.

Aarseth, Espen J. (1997). *Cybertext: Perspectives on Ergodic Literature*. Baltimore, MD: Johns Hopkins University Press.

Aarseth, Espen, and Stephan Günzel. (2019). "Introduction." In Espen Aarseth and Stephan Günzel (Eds.), *Ludotopia: Spaces, Places and Territories in Computer Games* (pp. 7–9). Bielefeld, Germany: Transcript.

Aerometrex. (2022). "How Digital Twins Are Transforming Our Cities." *Aerometrex Blog*. Accessed January 20, 2022. https://aerometrex.com.au/resources/blog/how-digital-twins-are-transforming-our-cities/.

AFIPS. (1962). "EDP as a Natural Resource." *Proceedings of the 1962 Fall Joint Computer Conference*, 71–72 (Philadelphia, December 1962). Washington, DC: Spartan Books.

Afonso, Nuno, and Rui Prada. (2008). "Agents That Relate: Improving the Social Believability of Non-Player Characters in Role-Playing Games." In Scott M. Stevens and Shirley J. Saldamarco (Eds.), *Entertainment Computing—ICEC 2008*. Lecture Notes in Computer Science 5309 (pp. 34–45). Berlin: Springer.

Alexander, Christopher, Sara Ishikawa, Murray Silverstein, Max Jacobson, Ingrid Fiksdahl King, and Shlomo Angel. (1977). *A Pattern Language: Towns, Buildings, Construction*. New York: Oxford University Press.

Alfrink, Kars. (2014). "The Gameful City." In Steffen P. Walz and Sebastian Deterding (Eds.), *The Gameful World: Approaches, Issues, Applications* (pp. 527–560). Cambridge, MA: MIT Press.

Ambinder, Mike. (2011). "Biofeedback in Gameplay: How Valve Measures Physiology to Enhance Gaming Experience." *Game Developers Conference* (March 3, 2011). Accessed November 30, 2021. www.gdcvault.com/play/1014734/Biofeedback-in-Gameplay-How-Valve.

Andrews, Chris. (2021). "ArcGIS: A Foundation for Digital Twins." *Esri ArcGIS Blog* (January 11, 2021). Accessed January 20, 2022. www.esri.com/arcgis-blog/products/arcgis/aec/gis-foundation-for-digital-twins/.

Ang, Carmen. (2020). "The Growth of Home Fitness Apps." *Visual Capitalist* (September 10, 2020). Accessed September 5, 2021. www.visualcapitalist.com/the-growth-of-home-fitness-apps-2020/.

Apple. (2020). "Exposure Risk Value Calculation in ExposureNotification Version 1." *Apple Developer Documentation.* Accessed September 5, 2021. https://developer.apple.com/documentation/exposurenotification/enexposureconfiguration/exposure_risk_value_calculation_in_exposurenotification_version_1.

Apprich, Clemens. (2018). "Secret Agents: A Psychoanalytic Critique of Artificial Intelligence and Machine Learning." *Digital Culture and Society* 4, no. 1: 29–44.

Arora, Neelima, Amit K. Banerjee, and Mangamoori L. Narasu. (2020). "The Role of Artificial Intelligence in Tackling COVID-19." *Future Virology* 15, no. 11 (November 2020): 717–724. Accessed September 5, 2021. www.ncbi.nlm.nih.gov/pmc/articles/PMC7692869/pdf/fvl-2020-0130.pdf.

Atherton, Kelsey. (2020). "DARPA Wants Wargame AI to Never Fight Fair." *Breaking Defense* (August 18, 2020). Accessed September 5, 2021. https://breakingdefense.com/2020/08/darpa-wants-wargame-ai-to-never-fight-fair/.

Ball, Matthew. (2020). "The Metaverse: What It Is, Where to Find It, Who Will Build It, and *Fortnite.*" *MatthewBall.vc* (January 13, 2020). Accessed October 26, 2021. www.matthewball.vc/all/themetaverse.

Ball, Matthew. (2021). "Framework for the Metaverse." *MatthewBall.vc* (June 29, 2021). Accessed October 26, 2021. www.matthewball.vc/all/forwardtothemetaverseprimer.

Banerjee, Shayak, David Stevens, and Santosh Thammana. (2021a). "How We Built: An Early-Stage Recommender System." *Peloton Press* (October 18, 2021). Accessed January 10, 2022. www.onepeloton.com/press/articles/designing-an-early-stage-recommender-system.

Banerjee, Shayak et al. (2021b). "Personalizing Peloton: Combining Rankers and Filters to Balance Engagement and Business Goals." *Proceedings of the Fifteenth ACM Conference on Recommender Systems (RecSys'21)*, 575–576 (Amsterdam, Netherlands, September 27–October 1, 2021). New York: Association for Computing Machinery.

Banet-Weiser, Sarah. (2012). "Branding the Crisis." In Manuel Castells, João Caraça, and Gustavo Cardoso (Eds.), *Aftermath: The Cultures of the Economic Crisis* (pp. 107–131). Oxford: Oxford University Press.

Baym, Nancy K. (1998). "The Emergence of On-Line Community." In Steven G. Jones (Ed.), *Cybersociety 2.0: Revisiting Computer-Mediated Communication and Community* (pp. 35–68). Thousand Oaks, CA: Sage Publications.

Belden, Thomas G. et al. (1961). *Computers in Command and Control.* Technical Report 61–12 (November 1961). Institute for Defense Analyses Research and Engineering Support Division. Arlington, VA: Armed Services Technical Information Agency. Accessed August 8, 2021. https://apps.dtic.mil/sti/pdfs/AD0271997.pdf.

Bellemare, Marc, Yavar Naddaf, Joel Veness, and Michael Bowling. (2013). "The Arcade Learning Environment: An Evaluation Platform for General Agents." *Journal of Artificial Intelligence Research* 47: 253–279.

Bellomy, Ian. (2017). "What Counts: Configuring the Human in Platform Studies." *Analog Game Studies* 4, no. 2 (March 20, 2017). Accessed June 12, 2018. https://analoggamestudies.org/2017/03/what-counts/.

Benjamin, Ruha. (2019). *Race After Technology: Abolitionist Tools for the New Jim Code.* Cambridge: Polity Press.

Beyvers, Sarah E. (2020). "The Game of Narrative Authority: Subversive Wandering and Unreliable Narration in *The Stanley Parable*." *Journal of Gaming & Virtual Worlds* 12, no. 1 (March 2020): 7–21.

Björk, Staffan, and Jussi Holopainen. (2006). "Games and Design Patterns." In Katie Salen and Eric Zimmerman (Eds.), *The Game Design Reader: A Rules of Play Anthology* (pp. 410–437). Cambridge, MA: MIT Press.

Bjørkelo, Kristian A. (2018). "'It Feels Real to Me': Transgressive Realism in *This War of Mine*." In Kristine Jørgensen and Faltin Karlsen (Eds.), *Transgression in Games and Play* (pp. 169–185). Cambridge, MA: MIT Press.

Bogost, Ian. (2007). *Persuasive Games: The Expressive Power of Videogames*. Cambridge, MA: MIT Press.

Bogost, Ian, Michael Mateas, Janet Murray, and Michael Nitsche. (2005). "Asking What Is Possible: The Georgia Tech Approach to Game Research and Education." *The International Digital Media and Arts Association Journal* 2, no. 1: 59–68.

Booth, Mike. (2009). "The AI Systems of *Left 4 Dead*." *Fifth Artificial Intelligence and Interactive Digital Entertainment Conference* (Stanford, CA, October 2009). Accessed August 8, 2021. https://steamcdn-a.akamaihd.net/apps/valve/2009/ai_systems_of_l4d_mike_booth.pdf.

Boschert, Stefan, and Roland Rosen. (2016). "Digital Twin—The Simulation Aspect." In Peter Hehenberger and David Bradley (Eds.), *Mechatronic Futures: Challenges and Solutions for Mechatronic Systems and Their Designers* (pp. 59–74). Cham, Switzerland: Springer.

Botta, Mark. (2015). "Infected AI in *The Last of Us*." In Steve Rabin (Ed.), *Game AI Pro 2: Collected Wisdom of Game AI Professionals* (pp. 407–418). Boca Raton, FL: CRC Press.

Bratton, Benjamin H. (2015). *The Stack: On Software and Sovereignty*. Cambridge, MA: MIT Press.

Burchell, Graham. (1993). "Liberal Government and Techniques of the Self." *Economy and Society* 22, no. 3 (August 1993): 267–282.

Bush, Vannevar. (1945). "As We May Think." *The Atlantic Monthly* (July 1945): 101–108.

Calvo, Rafael A., Sebastian Deterding, and Richard M. Ryan. (2020). "Health Surveillance During COVID-19 Pandemic: How to Safeguard Autonomy and Why It Matters." *BMJ* (April 6, 2020). Accessed September 5, 2021. www.bmj.com/content/bmj/369/bmj.m1373.full.pdf.

Campbell, Colin. (2014). "Ebola Scare Drives Sharp Rise in *Plague Inc.* Downloads." *Polygon* (October 27, 2014). Accessed May 6, 2020. www.polygon.com/2014/10/27/7080069/ebola-plague-game-spikes-in-popularity.

Cardoso, Gustavo, and Pedro Jacobetty. (2012). "Surfing the Crisis: Cultures of Belonging and Networked Social Change." In Manuel Castells, João Caraça, and Gustavo Cardoso (Eds.), *Aftermath: The Cultures of the Economic Crisis* (pp. 177–209). Oxford: Oxford University Press.

CD Projekt. (2016). "The CD Projekt Group Secures 30 Million PLN in GameINN Innovation Funds." *CD Projekt News* (December 15, 2016). Accessed November 18, 2021. www.cdprojekt.com/en/media/news/cd-projekt-group-secures-30-million-pln-gameinn-innovation-funds/.

CD Projekt. (2017). "1/2017 City Creation." *EU Projects* (March 14, 2017). Accessed November 18, 2021. www.cdprojekt.com/en/capital-group/eu-projects/12017-city-creation/.

Certeau, Michel de. (1984). *The Practice of Everyday Life*. Translated by Steven Rendall. Berkeley, CA: University of California Press.

Chang, Alenda Y. (2019). *Playing Nature: Ecology in Video Games*. Minneapolis: University of Minnesota Press.

Cheney-Lippold, John. (2011). "A New Algorithmic Identity: Soft Biopolitics and the Modulation of Control." *Theory, Culture and Society* 28, no. 6: 164–181.

Chew, Han Shi Jocelyn, Wei How Darryl Ang, and Ying Lau. (2021). "The Potential of Artificial Intelligence in Enhancing Adult Weight Loss: A Scoping Review." *Public Health Nutrition* 24, no. 8: 1993–2020.

Chia, Aleena, Brendan Keogh, Dale Leorke, and Benjamin Nicoll. (2020). "Platformisation in Game Development." *Internet Policy Review* 9, no. 4 (October 21, 2020): 1–28.

Chou, Allen. (2016). "A Brain Dump of What I Worked on for *Uncharted 4*." *AllenChou.net* (May 10, 2016). Accessed May 15, 2020. http://allenchou.net/2016/05/a-brain-dump-of-what-i-worked-on-for-uncharted-4/.

Chun, Wendy Hui Kyong. (2011). *Programmed Visions: Software and Memory*. Cambridge, MA: MIT Press.

Colman, Felicity, Vera Bühlmann, Aislinn O'Donnell, and Iris van der Tuin. (2018). "Ethics of Coding: A Report on the Algorithmic Condition." H2020-EU.2.1.1.— Industrial Leadership—Leadership in enabling and industrial technologies— Information and Communication Technologies. Brussels: European Commission (Project Number 732407). https://cordis.europa.eu/project/rcn/207025_en.html.

Colmerauer, Alain, and Philippe Roussel. (1993). "The Birth of Prolog." *ACM SIGPLAN Notices* 28, no. 3 (March 1993): 37–52.

Compaq Computer. (1996). "Internet Solutions Division Strategy for Cloud Computing." November 14, 1996. Accessed August 4, 2021. https://s3.amazonaws.com/files.technologyreview.com/p/pub/legacy/compaq_cst_1996_0.pdf.

Consalvo, Mia. (2007). *Cheating: Gaining Advantage in Videogames*. Cambridge, MA: MIT Press.

Cooley, Benjamin. (2021). "Data Visualization as Grief: What We Gain When We Think beyond Spike Maps, Choropleths, and Curvy Case Charts." *Medium* (January 22, 2021). Accessed January 27, 2021. https://medium.com/nightingale/data-visualization-as-grief-599dc6536e6f.

Crawford, Kate. (2021). *Atlas of AI: Power, Politics, and the Planetary Costs of Artificial Intelligence*. New Haven, CT: Yale University Press.

Crawford, Kate, and Ryan Calo. (2016). "There Is a Blind Spot in AI Research." *Nature* 538 (October 20, 2016): 311–313.

Crawford, Kate and Vladan Joler. (2018). "Anatomy of an AI System: The Amazon Echo as An Anatomical Map of Human Labor, Data and Planetary Resources." *AI Now Institute and Share Lab* (September 7, 2018). Accessed January 27, 2021. https://anatomyof.ai.

D'Anastasio, Cecilia. (2022). "Gaming Giant Unity Wants to Digitally Clone the World." *Wired* (January 18, 2022). Accessed January 28, 2022. www.wired.com/story/gaming-giant-unity-wants-to-digitally-clone-the-world/.

Daily Mail. (2008). "Obesity Experts Condemn Nintendo's Wii Fit Game after It Tells 10-year-old Girl She's Fat." May 16, 2008. Accessed January 15, 2022. www.dailymail.co.uk/news/article-566754/Obesity-experts-condemn-Nintendos-Wii-Fit-game-tells-10-year-old-girl-shes-fat.html.

Davis, Paul K. (1988). *Applying Artificial Intelligence Techniques to Strategic-Level Gaming and Simulation (N-2752-RC)*. Santa Monica, CA: RAND Corporation. Accessed January 5, 2022. www.rand.org/content/dam/rand/pubs/notes/2007/N2752.pdf.

De Monchaux, Nicholas. (2021). "The City and the City." *Wired* 29, no. 12 (December 2021–January 2022): 94–97.

De Plater, Michael et al. (2016). "Nemesis Characters, Nemesis Forts, Social Vendettas and Followers in Computer Games." *United States Patent Application Publication US 2016/0279522 A1* (September 29, 2016). Accessed September 18, 2021. https://patentimages.storage.googleapis.com/e5/a5/24/5d3b11e0fc0d98/US20160279522A1.pdf.

Defense Advanced Research Projects Agency (DARPA). (1983). *Strategic Computing: New-Generation Computing Technology: A Strategic Plan for its Development and Application to Critical Problems in Defense.* October 28, 1983. Washington, DC: United States Department of Defense. Accessed August 7, 2021. www.nitrd.gov/nitrdgroups/images/3/3a/20040929_strategic_computing.pdf.

Defense Advanced Research Projects Agency (DARPA). (2020). "Gamebreak AI Effort Gets Under Way." *DARPA: News* (May 13, 2020). Accessed January 5, 2022. www.darpa.mil/news-events/2020-05-13.

Department for Business, Energy and Industrial Strategy. (2017). *Industrial Strategy: Building a Britain Fit for the Future.* November 27, 2017. Accessed August 15, 2021. https://assets.publishing.service.gov.uk/government/uploads/system/uploads/attachment_data/file/664563/industrial-strategy-white-paper-web-ready-version.pdf.

Deterding, Sebastian, Dan Dixon, Rilla Khaled, and Lennart Nacke. (2011). "From Game Design Elements to Gamefulness: Defining 'Gamification'." *Proceedings of the Fifteenth International Academic MindTrek Conference: Envisioning Future Media Environments,* 9–15 (Tampere, Finland, September 2011). New York: Association for Computing Machinery.

Digital Domain. (2020). "Digital Domain Announces Masquerade 2.0—Blockbuster Facial Capture for Next-Gen Games." *Digital Domain News.* Accessed January 5, 2022. https://digitaldomain.com/news/digital-domain-announces-masquerade-2-0-blockbuster-facial-capture-for-next-gen-games/.

Dill, Kevin. (2014). "What Is Game AI?" In Steven Rabin (Ed.), *Game AI Pro: Collected Wisdom of Game AI Professionals* (pp. 3–9). Boca Raton, FL: CRC Press.

Dourish, Paul. (2017). *The Stuff of Bits: An Essay on the Materialities of Information.* Cambridge, MA: MIT Press.

Dourish, Paul, and Melissa Mazmanian. (2013). "Media as Material: Information Representations as Material Foundations for Organizational Practice." (Working paper for Third International Symposium on Process Organization Studies. (Corfu, Greece, 16–18 June 2011)) In Paul R. Carlile, Davide Nicolini, Ann Langley, and Haridimos Tsoukas (Eds.), *How Matter Matters: Objects, Artifacts, and Materiality in Organization Studies* (pp. 92–118). Oxford, UK: Oxford University Press.

Dyckhoff, Max. (2015). "Ellie: Buddy AI in *The Last of Us*." In Steve Rabin (Ed.), *Game AI Pro 2: Collected Wisdom of Game AI Professionals* (pp. 431–442). Boca Raton, FL: CRC Press.

Ebadi, Salehe Erfanian et al. (2021). *PeopleSansPeople: A Synthetic Data Generator for Human-Centric Computer Vision.* arXiv:2112.09290 (December 17, 2021). Accessed January 5, 2022. https://arxiv.org/pdf/2112.09290.pdf.

Ecoffet, Adrien, Joost Huizinga, Joel Lehman, Kenneth Stanley, and Jeff Clune. (2021). "First Return, Then Explore." *Nature* 590 (February 24, 2021): 580–586.

Ellerbrok, Ariane. (2011). "Playful Biometrics: Controversial Technology Through the Lens of Play." *The Sociological Quarterly* 52, no. 4 (Fall 2011): 528–547.

Elleström, Lars. (2020). "Transmediation: Some Theoretical Considerations." In Niklas Salmose and Lars Elleström (Eds.), *Transmediations: Communication across Media Borders* (pp. 1–14). New York: Routledge.

Elliott, Anthony. (2019). *The Culture of AI: Everyday Life and the Digital Revolution.* New York: Routledge.

Elsami, S. M. Ali et al. (2018). "Neural Scene Representation and Rendering." *Science* 360, no. 6394 (June 15, 2018): 1204–1210.

Engelbart, Douglas C. (1962). *Augmenting Human Intellect: A Conceptual Framework.* Summary Report AFOSR-3233 (October 1962). Menlo Park, CA: Stanford Research Institute.

Epic Games. (2021). "What Balenciaga's *Afterworld: The Age of Tomorrow* Tells Us About the Future of Fashion." *Epic Games Unreal Engine Spotlights.* Accessed February 14, 2022. www.unrealengine.com/en-US/spotlights/what-balenciaga-s-afterworld-the-age-of-tomorrow-tells-us-about-the-future-of-fashion.

Epic Games. (2022). *MetaHuman Creator End User License Agreement.* Accessed February 14, 2022. www.unrealengine.com/en-US/eula/mhc.

European Commission. (2021). *Destination Earth.* Brochure (March 19, 2021). Accessed February 15, 2022. https://digital-strategy.ec.europa.eu/en/library/destination-earth.

Evans, Richard. (2002). "Varieties of Learning." In Steve Rabin (Ed.), *AI Game Programming Wisdom* (pp. 567–578). Hingham, MA: Charles River Media.

Farca, Gerald. (2016). "The Emancipated Player." *DiGRA/FDG'16—Proceedings of the First International Joint Conference of DiGRA and FDG* 13, no. 1 (August 2016). Accessed March 21, 2021. www.digra.org/wp-content/uploads/digital-library/paper_205.pdf.

Farca, Gerald, and Charlotte Ladevèze. (2016). "The Journey to Nature: *The Last of Us* as Critical Dystopia." *DiGRA/FDG'16—Proceedings of the First International Joint Conference of DiGRA and FDG* 13, no. 1 (August 2016). Accessed March 21, 2021. www.digra.org/wp-content/uploads/digital-library/paper_246.pdf.

Feigenbaum, Edward, Pamela McCorduck, and H. Penny Nii. (1988). *The Rise of the Expert Company: How Visionary Companies are Using Artificial Intelligence to Achieve Higher Productivity and Profits.* New York: Times Books.

Felleisen, Matthias. (1991). "On the Expressive Power of Programming Languages." *Science of Computer Programming* 17, no. 1–3 (December 1991): 35–75.

Ferguson, Stephen. (2020). "Apollo 13: The First Digital Twin." *Siemens Blog* (April 14, 2020). Accessed January 20, 2022. https://blogs.sw.siemens.com/simcenter/apollo-13-the-first-digital-twin/.

Fink, Charlie. (2018). "6D.ai Funded to Build the AR Cloud." *Forbes* (March 29, 2018). Accessed January 5, 2022. www.forbes.com/sites/charliefink/2018/03/29/6d-ai-funded-to-build-the-ar-cloud/?sh=80d801b6bb60.

Finn, Ed. (2017). *What Algorithms Want: Imagination in the Age of Computing.* Cambridge, MA: MIT Press.

Fizek, Sonia. (2014). "Why Fun Matters: In Search of Emergent Playful Experiences." In Mathias Fuchs, Sonia Fizek, Paolo Ruffino, and Niklas Schrape (Eds.), *Rethinking Gamification* (pp. 273–287). Lüneberg, Germany: Meson Press.

Fizek, Sonia. (2018). "Automated State of Play: Rethinking Anthropocentric Rules of the Game." *Digital Culture and Society* 4, no. 1: 201–214.

Flyverbom, Mikkel, Anders Koed Madsen, and Andreas Rasche. (2017). "Big Data as Governmentality: Digital Traces, Algorithms, and the Reconfiguration of Data for International Development." *Information Society* 33, no. 1: 35–42.

Fogel, David. (2005). "Preface." In Graham Kendall and Simon Lucas (Eds.), *Proceedings of the IEEE 2005 Symposium on Computational Intelligence and Games,* 5–6 (Colchester,

Essex, UK, April 2005). Accessed November 21, 2021. https://citeseerx.ist.psu.edu/viewdoc/download?doi=10.1.1.728.5120&rep=rep1&type=pdf.

Foley, John Paul et al. (2015). "Exercise System and Method." *United States Patent Number US 9,174,085 B2* (November 3, 2015). Accessed January 15, 2022. https://patent images.storage.googleapis.com/84/24/c7/4730b0b1d80230/US9174085.pdf.

Foucault, Michel. (1986). "Of Other Spaces." Translated by Jay Miskowiec. *Diacritics* 16, no. 1 (Spring 1986): 22–27.

Foucault, Michel. (1991): "Governmentality." In Graham Burchell, Colin Gordon, and Peter Miller (Eds.), *The Foucault Effect: Studies in Governmentality* (pp. 87–104). Chicago: University of Chicago Press.

Freedman, Eric. (2011). *Transient Images: Personal Media in Public Frameworks*. Philadelphia, PA: Temple University Press.

Freedman, Eric. (2020). *The Persistence of Code in Game Engine Culture*. New York: Routledge.

Freeman-Mills, Max. (2018). "Mike Booth, The Architect of *Left 4 Dead's* AI Director, Explains Why It's So Bloody Good." *Kotaku* (November 21, 2018). Accessed May 15, 2020. www.kotaku.com.au/2018/11/mike-booth-the-architect-of-left-4-deads-ai-director-explains-why-its-so-bloody-good/.

Fuchs, Mathias, and Andreas Sudmann. (2019). "Games and AI: Paths, Challenges, Critique." *Eludamos. Journal for Computer Game Culture* 10, no. 1: 1–7.

Gallant, Matthew. (2017). "Authored vs. Systemic: Finding a Balance for Combat AI in *Uncharted 4*." *Game Developers Conference* (March 2017). Accessed May 15, 2020. www.gdcvault.com/play/1024467/Authored-vs-Systemic-Finding-a.

GamesIndustry.biz. (2017). "Creative Assembly: Early Stage Specialism Is What Studios Look for." *GamesIndustry.biz* (September 6, 2017). Accessed November 18, 2021. www.gamesindustry.biz/articles/2017-09-06-creative-assembly-specialising-at-early-stage-is-what-studios-look-for.

Gelernter, David. (1991). *Mirror Worlds: Or the Day Software Puts the Universe in a Shoebox . . . How It Will Happen and What It Will Mean*. New York: Oxford University Press.

Génération Robots. (2021). "Payload: Boston Dynamics Spot CORE AI." Accessed July 28, 2021. www.generationrobots.com/en/403769-payload-boston-dynamics-spot-core-ai.html.

Georgeson, Jeffrey, and Christopher Child. (2016). "NPCs as People, Too: The Extreme AI Personality Engine." Accessed March 30, 2021. https://arxiv.org/ftp/arxiv/papers/1609/1609.04879.pdf.

Gerlach, Neil, and Sheryl Hamilton. (2014). "Trafficking in the Zombie: The CDC Zombie Apocalypse Campaign, Diseaseability and Pandemic Culture." *Refractory: A Journal of Entertainment Media* 23 (June 2014). Accessed May 6, 2020. http://refractory.unimelb.edu.au/2014/06/26/cdc-zombie-apocalypse-gerlach-hamilton/.

Gevarter, William B. (1982). *An Overview of Expert Systems*. NBSIR 82–2505. May 1982. Washington, DC: United States Department of Commerce. Accessed August 7, 2021. www.govinfo.gov/content/pkg/GOVPUB-C13-317d16eca06b44face3805d30 021d02d/pdf/GOVPUB-C13-317d16eca06b44face3805d30021d02d.pdf.

Giddings, Seth. (2005). "Playing with Non-Humans: Digital Games as Techno-Cultural Form." *Proceedings of the 2005 DiGRA International Conference: Changing Views: Worlds in Play* (Vancouver, British Columbia, Canada, June 16–20, 2005). Accessed August 13, 2021. www.digra.org/digital-library/publications/playing-with-non-humans-digital-games-as-techno-cultural-form/.

Gillespie, Tarleton. (2010). "The Politics of 'Platforms.'" *New Media and Society* 12, no. 3: 347–364.

Gillespie, Tarleton. (2014). "The Relevance of Algorithms." In Tarleton Gillespie, Pablo J. Boczkowski, and Kirsten A. Foot (Eds.), *Media Technologies: Essays on Communication, Materiality, and Society* (pp. 167–193). Cambridge, MA: MIT Press.

Glas, René. (2015). "Of Heroes and Henchmen: The Conventions of Killing Generic Expendables in Digital Games." In Torill Elvira Mortensen, Jonas Linderoth, and Ashley M. L. Brown (Eds.), *The Dark Side of Game Play: Controversial Issues in Playful Environments* (pp. 33–49). New York: Routledge.

Goffey, Andrew. (2008). "Algorithm." In Matthew Fuller (Ed.), *Software Studies: A Lexicon* (pp. 15–20). Cambridge, MA: MIT Press.

Goffey, Andrew. (2014). "Technology, Logistics and Logic: Rethinking the Problem of Fun in Software." In Olga Goriunova (Ed.), *Fun and Software: Exploring Pleasure, Paradox and Pain in Computing* (pp. 21–40). New York: Bloomsbury.

Golumbia, David. (2009). *The Cultural Logic of Computation.* Cambridge, MA: Harvard University Press.

González-Hermida, Martín et al. (2019). "Study of Artificial Intelligent Algorithms Applied in Procedural Content Generation in Video Games." *Eludamos. Journal for Computer Game Culture* 10, no. 1: 39–54.

Gordon, Bing. (2005). "The Turing Test for Game AI" [PowerPoint slides]. *Proceedings of the First Artificial Intelligence and Interactive Digital Entertainment Conference* (Marina del Rey, CA, June 2005). Menlo Park, CA: AAAI Press. Accessed November 20, 2021. www.aaai.org/Library/AIIDE/aiide05contents.php.

Gordon, Eric, and Gabriel Mugar. (2020). *Meaningful Inefficiencies: Civic Design in an Age of Digital Expediency.* New York: Oxford University Press.

Greenberger, Martin. (1964). "The Computers of Tomorrow." *The Atlantic Monthly* (May 1964). Accessed August 6, 2021. www.tnellen.com/cybereng/ebooks/greenbf. htm.

Greenfield, Adam. (2017). *Radical Technologies: The Design of Everyday Life.* Brooklyn, NY: Verso.

Gregory, Jason. (2009). *Game Engine Architecture.* Natick, MA: A K Peters.

Gregory, Jason. (2014). "Context-Aware Dialog in *The Last of Us.*" *Game Developers Conference* (March 2014). Accessed May 15, 2020. www.youtube.com/watch?v=Y7-OoXqNYgY.

Grieves, Michael. (2005). "Product Lifecycle Management: The New Paradigm for Enterprises." *International Journal of Product Development* 2, no. 1–2: 71–84.

Grieves, Michael, and John Vickers. (2017). "Digital Twin: Mitigating Unpredictable, Undesirable Emergent Behavior in Complex Systems." In Franz-Josef Kahlen, Shannon Flumerfelt, and Anabela Alves (Eds.), *Transdisciplinary Perspectives on Complex Systems: New Findings and Approaches* (pp. 85–113). Cham, Switzerland: Springer.

Gröger, Christoph. (2021). "There Is No AI Without Data." *Communications of the ACM* 64, no. 11 (November 2021): 98–108.

Harlow, Francis H., and Nicholas Metropolis. (1983). "Weapons Simulation Leads to the Computer Era." *Los Alamos Science* (Winter/Spring 1983): 132–141.

Hayles, N. Katherine. (2012). *How We Think: Digital Media and Contemporary Technogenesis.* Chicago: University of Chicago Press.

Henricks, Thomas S. (2014). "Play as Self-Realization: Toward a General Theory of Play." *American Journal of Play* 6, no. 2 (Winter 2014): 190–213.

Hernández-Orallo, José, and Karina Vold. (2019). "AI Extenders: The Ethical and Societal Implications of Humans Cognitively Extended by AI." *Proceedings of the 2019 AAAI/ ACM Conference on AI, Ethics, and Society*, 507–513 (Honolulu, HI, January 2019). New York: Association for Computing Machinery.

Hicks, Ben. (2019). "Industry 4.0 and Digital Twins: Key Lessons From NASA." *The Future Factory Blog*. Accessed January 20, 2022. www.thefuturefactory.com/blog/24.

Hong, Sun-ha. (2020). *Technologies of Speculation: The Limits of Knowledge in a Data-Driven Society*. New York: New York University Press.

Hsu, Hansen. (2020). "AI and Play, Part 1: How Games Have Driven Two Schools of AI Research." *Computer History Museum Blog* (July 23, 2020). Accessed March 28, 2021. https://computerhistory.org/blog/ai-and-play-part-1-how-games-have-driven-two-schools-of-ai-research/.

Hudlicka, Eva. (2009). "Affective Game Engines: Motivation and Requirements." *FDG'09: Proceedings of the Fourth International Conference on Foundations of Digital Games*, 299–306 (Orlando, FL, April 2009). New York: Association for Computing Machinery.

Huizinga, Johan. (1949). *Homo Ludens: A Study of the Play-Element in Culture*. London: Routledge & Kegan Paul.

Hunicke, Robin, Marc LeBlanc, and Robert Zubek. (2004). "MDA: A Formal Approach to Game Design and Game Research." *Proceedings of the Challenges in Game AI Workshop, Nineteenth National Conference on Artificial Intelligence* (San Jose, CA, 2004). Accessed August 9, 2009. http://users.cs.northwestern.edu/~hunicke/pubs/MDA.pdf.

Hyundai Motor Company. (2016). "2030 Megatrend: 12 Megatrends that are Most Likely to Affect Car Industry in 2030. Primary Output of Project IONIQ Lab." *Hyundai: Newsroom* (July 22, 2016). Accessed July 28, 2021. www.hyundai.news/eu/articles/press-releases/project-ioniq-lab-to-drive-future-mobility-innovation.html.

Hyundai Motor Company. (2021). "Hyundai x Boston Dynamics: Welcome to the Future of Mobility." *Hyundai: Brand*. Accessed July 17, 2021. www.hyundai.com/worldwide/en/brand/hyundai-boston-dynamics.

Hyundai Motor Company. (2022). "Hyundai Motor and Unity to Build Meta-Factory Accelerating Intelligent Manufacturing Innovation." *Hyundai Motor Group Newsroom* (January 7, 2022). Accessed March 7, 2022. www.prnewswire.com/news-releases/hyundai-motor-and-unity-partner-to-build-meta-factory-accelerating-intelligent-manufacturing-innovation-301455322.html.

IBM. (2021). "AI for Kids Catalog." *Data and AI Learning Catalog*. Accessed July 18, 2021. http://ibmtvdemo.edgesuite.net/software/DAELL/Test/POC/ai4kids.html.

IBM. (2021). "IBM Process Mining." *IBM Cloud Pak for Business Automation*. Accessed January 29, 2022. www.ibm.com/cloud/cloud-pak-for-business-automation/process-mining?utm_content=SRCWW&p1=Search&p4 = 43700065994437098&p5=p&gclid=EAIaIQobChMI4fDmwObX9QIVR2xvBB2EywptEAAYASAAEgLuPvD_BwE&gclsrc=aw.ds.

ICON Health & Fitness, Inc. v. Peloton Interactive, Inc., 1:21-cv-00507, No. 1 (District of Delaware April 7, 2021). Accessed January 15, 2022. www.docketalarm.com/cases/Delaware_District_Court/1-21-cv-00507/ICON_Health_&_Fitness_Inc._v._Peloton_Interactive_Inc/1/.

iFIT. (2021). *iFIT: Our Story*. Accessed January 15, 2022. https://company.ifit.com/en/our-story/.

IHRSA. (2021). *2021 IHRSA Media Report: Health and Fitness Consumer Data and Industry Trends*. Accessed January 15, 2022. www.ihrsa.org/publications/2021-ihrsa-media-report-2/.

Isla, Damian. (2005). "GDC 2005 Proceeding: Handling Complexity in the *Halo 2* AI." *Game Developer* (March 10, 2005). Accessed November 20, 2021. www.gamedeveloper. com/programming/gdc-2005-proceeding-handling-complexity-in-the-i-halo-2-i-ai.

Jacobs, Jane. (1961). *The Death and Life of Great American Cities*. New York: Vintage Books.

Johnson, Robin. (2014). "Artificial Intelligence." In Mark J. P. Wolf and Bernard Perron (Eds.), *The Routledge Companion to Video Game Studies* (pp. 10–18). New York: Routledge.

Juliani, Arthur et al. (2020). "Unity: A General Platform for Intelligent Agents." arXiv:1809.02627v2 [cs.LG]. Accessed March 21, 2021. https://arxiv.org/abs/1809. 02627v2.

Justesen, Niels et al. (2017). "Deep Learning for Video Game Playing." arXiv:1708.07902v3 [cs.AI]. Accessed March 21, 2021. https://arxiv.org/abs/1708.07902v3.

Juul, Jesper. (2005). *Half-Real: Video Games between Real Rules and Fictional Worlds*. Cambridge, MA: MIT Press.

Juul, Jesper. (2017). "The Darkening of Play." *The Ludologist* blog (January 19, 2017). Accessed March 21, 2021. www.jesperjuul.net/ludologist/2017/01/19/the-darkening-of-play/.

Kamalzadeh, Mohsen. (2021). "Unlocking Intelligent Solutions in the Home with Computer Vision." *Unity Blog* (September 21, 2021). Accessed January 5, 2022. https://blog.unity.com/ technology/unlocking-intelligent-solutions-in-the-home-with-computer-vision.

Karras, Tero, Samuli Laine, and Timo Aila. (2019). "A Style-Based Generator Architecture for Generative Adversarial Networks." *Proceedings of the IEEE/CVF Conference on Computer Vision and Pattern Recognition* (June 2019): 4396–4405.

Keogh, Brendan. (2018). *A Play of Bodies: How We Perceive Videogames*. Cambridge, MA: MIT Press.

Khan, Ali S. (2011). "Preparedness 101: Zombie Apocalypse." *Centers for Disease Control and Prevention: Public Health Matters Blog* (May 16, 2011). Accessed March 22, 2021. https:// blogs.cdc.gov/publichealthmatters/2011/05/preparedness-101-zombie-apocalypse/.

Khemasuwan, Danai, and Henri G. Colt. (2021). "Applications and Challenges of AI-based Algorithms in the COVID-19 Pandemic." *BMJ Innovations* 7, no. 2 (April 2021): 387–398. Accessed September 5, 2021. https://innovations.bmj.com/content/bmjin-nov/7/2/387.full.pdf.

Khreiche, Mario. (2019). "Gamified Flow and the Sociotechnical Production of AI." *Eludamos. Journal for Computer Game Culture* 10, no. 1: 55–65.

Kirkpatrick, Graeme. (2013). *Computer Games and the Social Imaginary*. Cambridge: Polity Press.

Kister, James et al. (1957). "Experiments in Chess." *Journal of the Association for Computing Machinery* 4, no. 2 (April 1957): 174–177.

Kita, Chigusa Ishikawa. (2003). "J. C. R. Licklider's Vision for the IPTO." *IEEE Annals of the History of Computing* 25, no. 3 (July–September 2003): 62–77.

Kitchin, Rob, and Martin Dodge. (2011). *Code/Space: Software and Everyday Life*. Cambridge, MA: MIT Press.

Kolbe, Thomas H., and Andreas Donaubauer. (2021). "Semantic 3D City Modeling and BIM." In Wenzhong Shi et al. (Eds.), *Urban Informatics* (pp. 609–636). Singapore: Springer.

Kolko, Jon. (2010). *Thoughts on Interaction Design: A Collection of Reflections*. Burlington, MA: Morgan Kaufmann.

Korinek, Anton, and Joseph E. Stiglitz. (2021). "Covid-19 Driven Advances in Automation and Artificial Intelligence Risk Exacerbating Economic Inequality." *BMJ* 372, no. 367: 1–3.

Kuperman, Greg. (2020). "Gamebreaker." *Defense Advanced Research Projects Agency: Our Research*. Accessed January 5, 2022. www.darpa.mil/program/gamebreaker.

Laird, John, and Michael van Lent. (2001). "Human-Level AI's Killer Application: Interactive Computer Games." *AI Magazine* 22, no. 2 (Summer 2001): 15–25.

Lankoski, Petri, and Staffan Björk. (2007). "Gameplay Design Patterns for Believable Non-Player Characters." *DiGRA '07—Proceedings of the 2007 DiGRA International Conference: Situated Play* 4 (September 2007): 416–423. Accessed September 4, 2021. www. digra.org/wp-content/uploads/digital-library/07315.46085.pdf.

Latour, Bruno. (1990). "Technology is Society Made Durable." *The Sociological Review* 38, no. 1 (May 1990): 103–131.

Lefebvre, Henri. (1991). *The Production of Space*. Translated by Donald Nicholson-Smith. Malden, MA: Blackwell Publishing.

Leith, Douglas J., and Stephen Farrell. (2021). "Measurement-based Evaluation of Google/ Apple Exposure Notification API for Proximity Detection in a Commuter Bus." *PLOS ONE* 16, no. 4 (April 29, 2021): 1–16. Accessed September 5, 2021. https://journals. plos.org/plosone/article?id=10.1371/journal.pone.0250826.

Leslie, David. (2020). "Tackling COVID-19 through Responsible AI Innovation: Five Steps in the Right Direction." *Harvard Data Science Review* Special Issue 1 (June 5, 2020). Accessed September 5, 2021. https://hdsr.mitpress.mit.edu/pub/as1p81um/rel ease/3?readingCollection=0181d53b.

Lessig, Lawrence. (2006). *Code: Version 2.0*. New York: Basic Books.

Licklider, Joseph C. R. (1960). "Man-Computer Symbiosis." *IRE Transactions on Human Factors in Electronics* (March 1960): 4–11.

Linneman, John. (2016). "Tech Analysis: *Uncharted 4: A Thief's End*." *Eurogamer* (September 13, 2016). Accessed July 8, 2020. www.eurogamer.net/articles/ digitalfoundry-2016-uncharted-4-thiefs-end-tech-analysis.

Lucas, Simon et al. (2013). "Preface." In Simon Lucas, Michael Mateas, Mike Preuss, Pieter Spronck, and Julian Togelius (Eds.), *Artificial and Computational Intelligence in Games: A Follow-up to Dagstuhl Seminar 12191* (pp. vii–xi). Saarbrücken/Wadern, Germany: Dagstuhl Publishing.

Luciano, Dana, and Mel Y. Chen. (2015). "Has the Queer Ever Been Human?" *GLQ: A Journal of Lesbian and Gay Studies* 21, no. 2–3: 183–207.

Lululemon. (2020). "Mirror: This Is Not Just a Mirror. It's a Reflection of an Unstoppable Community." *Facebook* (September 28, 2020). Accessed May 7, 2022. https://www. facebook.com/watch/?v=786048188896731.

Lupton, Deborah. (2016). *The Quantified Self: A Sociology of Self-Tracking*. Malden, MA: Polity Press.

Mackenzie, Adrian. (2006). *Cutting Code: Software and Sociality*. New York: Peter Lang.

Makai, Péter Kristóf. (2020). "Three Ways of Transmediating a Theme Park: Spatializing Storyworlds in *Epic Mickey*, the *Monkey Island* Series and Theme Park Management Simulators." In Niklas Salmose and Lars Elleström (Eds.), *Transmediations: Communication Across Media Borders* (pp. 164–185). New York: Routledge.

Manovich, Lev. (2018). "Can We Think Without Categories?" *Digital Culture and Society* 4, no. 1: 17–27.

Martini, Annaclaudia, and Dorina Maria Buda. (2020). "Dark Tourism and Affect: Framing Places of Death and Disaster." *Current Issues in Tourism* 23, no. 6: 679–692.

Massachusetts Institute of Technology. (1961). "Management and the Computer of the Future." *Office of Public Relations* (April 28, 1961). Accessed August 4, 2021. https://

libraries.mit.edu/app/dissemination/DIPonline/AC0069_NewReleases/NewRe-
lease_1960/AC0069_1961/AC0069_196104_014.pdf.

Mateas, Michael. (2003a). "Expressive AI: Games and Artificial Intelligence." *DiGRA'03—
Proceedings of the 2003 DiGRA International Conference: Level Up* 2. Accessed March 21,
2021. www.digra.org/wp-content/uploads/digital-library/05150.02104.pdf.

Mateas, Michael. (2003b). "Expressive AI: A Semiotic Analysis of Machinic Affordances."
Proceedings of the Third Conference on Computational Semiotics for Games and New Media
(University of Teesside, UK, September 10–12, 2003), 58–67.

Mateas, Michael, and Noah Wardrip-Fruin. (2009). "Defining Operational Logics."
*DiGRA'09—Proceedings of the 2009 DiGRA International Conference: Breaking New
Ground: Innovation in Games, Play, Practice and Theory* 5 (September 2009). Accessed
March 27, 2021. www.digra.org/wp-content/uploads/digital-library/09287.21197.pdf.

Mäyrä, Frans. (2008). *An Introduction to Games Studies: Games in Culture*. London: Sage.

McCarthy, Anna. (2001). *Ambient Television: Visual Culture and Public Space*. Durham, NC:
Duke University Press.

McCarthy, John. (1959). "Programs with Common Sense." *Proceedings of a Symposium on
the Mechanisation of Thought Processes* (Volume 1), 75–84 (National Physical Laboratory,
Teddington, England, November 24–27, 1958). London: Her Majesty's Stationery
Office.

McCarthy, John. (1962). "Time-Sharing Computer Systems." In Martin Greenberger
(Ed.), *Computers and the World of the Future* (pp. 220–248). Cambridge, MA: MIT Press.

McCarthy, John et al. (1955). "A Proposal for the Dartmouth Summer Research Project
on Artificial Intelligence." (August 31, 1955). Accessed May 20, 2021. http://jmc.
stanford.edu/articles/dartmouth/dartmouth.pdf.

McCorduck, Pamela. (1979). *Machines Who Think: A Personal Inquiry into the History and
Prospects of Artificial Intelligence*. San Francisco: W. H. Freeman.

McGee, Kevin, and Aswin Thomas Abraham. (2010). "Real-Time Team-Mate AI in
Games: A Definition, Survey, and Critique." *FDG'10: Proceedings of the Fifth International
Conference on the Foundations of Digital Games*, 124–131 (Monterey, CA, June 2010).
New York: Association for Computing Machinery. Accessed September 28, 2021.
https://dl.acm.org/doi/pdf/10.1145/1822348.1822365.

McIntosh, Travis. (2015). "Human Enemy AI in *The Last of Us*." In Steve Rabin (Ed.),
Game AI Pro 2: Collected Wisdom of Game AI Professionals (pp. 419–429). Boca Raton,
FL: CRC Press.

McKeon, Albert. (2020). "Can Artificial Intelligence Apply Gaming to Military Strategy?"
Northrop Grumman: What We Do. Accessed January 5, 2022. www.northropgrumman.
com/what-we-do/can-artificial-intelligence-apply-gaming-to-military-strategy/.

Mehl, Lucien. (1959). "Automation in the Legal World: From the Machine Process-
ing of Legal Information to the 'Law Machine'." *Proceedings of a Symposium on the
Mechanisation of Thought Processes* (Volume 1), 757–779 (National Physical Laboratory,
Teddington, England, November 24–27, 1958). London: Her Majesty's Stationery
Office.

Metcalfe, Bob. (1973). "Ether Acquisition." (Memorandum). Xerox (May 22, 1973). Accessed
August 4, 2021. https://broadbandlibrary.com/bob-metcalfe-lays-down-the-law/.

Miikkulainen, Risto et al. (2006). "Computational Intelligence in Games." In Gary Yen
and David Fogel (Eds.), *Computational Intelligence: Principles and Practice* (pp. 155–191).
Piscataway, NJ: IEEE Computational Intelligence Society.

Millington, Ian, and John Funge. (2009). *Artificial Intelligence for Games (Second Edition)*. Burlington, MA: Elsevier.

Minsky, Marvin L. (1959). "Some Methods of Artificial Intelligence and Heuristic Programming." *Proceedings of a Symposium on the Mechanisation of Thought Processes* (Volume 1), 5–27 (National Physical Laboratory, Teddington, England, November 24–27, 1958). London: Her Majesty's Stationery Office.

Mirowski, Philip. (2003). "McCorduck's *Machines Who Think* after Twenty-Five Years: Revisiting the Origins of AI." *AI Magazine* 24, no. 4 (Winter 2003): 135–138.

Mitchell, Scott, and Sheryl Hamilton. (2018). "Playing at Apocalypse: Reading *Plague Inc.* in Pandemic Culture." *Convergence: The International Journal of Research into New Media Technologies* 24, no. 6 (December 2018): 587–606.

Mnih, Volodymyr et al. (2013). "Playing Atari with Deep Reinforcement Learning." *DeepMind* (January 2013). Accessed March 28, 2021. https://arxiv.org/pdf/1312.560 2v1.pdf.

Molino, Piero, Yaroslav Dudin, and Sai Sumanth Miryala. (2019). "Introducing Ludwig, a Code-Free Deep Learning Toolbox." *Uber Engineering Blog* (February 11, 2019). Accessed September 28, 2021. https://eng.uber.com/introducing-ludwig/.

Monsone, Cristina Rosaria, Eunika Mercier-Laurent, and Jósvai János. (2019). "The Overview of Digital Twins in Industry 4.0: Managing the Whole Ecosystem." *Proceedings of the Eleventh International Joint Conference on Knowledge Discovery, Knowledge Engineering and Knowledge Management (IC3K 2019)*, 271–276 (Vienna, Austria, September 17–19, 2019). Accessed January 20, 2022. www.scitepress.org/Papers/2019/83482/83482.pdf.

Moore, Phoebe V., and Jamie Woodcock. (2021). "AI: Making it, Faking it, Breaking it." In Phoebe V. Moore and Jamie Woodcock (Eds.), *Augmented Exploitation: Artificial Intelligence, Automation and Work* (pp. 1–9). London: Pluto Press.

Möring, Sebastian. (2019). "Distance and Fear: Defining the Play Space." In Espen Aarseth and Stephan Günzel (Eds.), *Ludotopia: Spaces, Places and Territories in Computer Games* (pp. 231–244). Bielefeld, Germany: Transcript.

Moto-oka, T. et al. (1981). "Challenge for Knowledge Information Processing Systems (Preliminary Report on Fifth Generation Computer Systems." *Proceedings of International Conference on Fifth Generation Computer Systems*, 1–85 (Tokyo, Japan, October 19–22, 1981). Tokyo: Japan Information Processing Development Center. Accessed August 7, 2021. www.jipdec.or.jp/archives/publications/J0002118.

Mozur, Paul, Raymond Zhong, and Aaron Krolik. (2020). "In Coronavirus Fight, China Gives Citizens a Color Code, With Red Flags." *New York Times* (March 1, 2020). Accessed September 5, 2021. www.nytimes.com/2020/03/01/business/china-corona virus-surveillance.html.

Murray, Janet. (1997). *Hamlet on the Holodeck: The Future of Narrative in Cyberspace*. New York: The Free Press.

National Center for Research and Development (NCBR). (2016). "List of Assessed Projects Submitted Under the Intelligent Development Operational Program 2014–2020, Action 1.2." Accessed November 18, 2021. https://archiwum.ncbr.gov.pl/fileadmin/gfx/ncbir/userfiles/_public/fundusze_europejskie/inteligentny_rozwoj/gameinn/lista_rankingowa_3_1.2_2016_poir__gameinn.pdf.

Navarro-Remesal, Victor. (2016). "Regarding the (Game) Pain of Others: Suffering and Compassion in Video Games." Paper Presented in Conference: *Concerns about Video Games and the Video Games of Concern*. Copenhagen: IT University.

New South Wales Government. (2022). "Spatial Digital Twin." *New South Wales Government Spatial Services: Projects*. Accessed January 20, 2022. www.spatial.nsw.gov.au/what_we_do/projects/digital_twin.

Newell, Allen, John Clifford Shaw, and Herbert A. Simon. (1957). "Empirical Explorations of the Logic Theory Machine: A Case Study in Heuristic." *Proceedings of the Western Joint Computer Conference*, 218–230 (Los Angeles, February 1957). New York: Association for Computing Machinery.

Newell, Allen, John Clifford Shaw, and Herbert A. Simon. (1963). "Chess-Playing Programs and the Problem of Complexity." In Edward A. Feigenbaum and Julian Feldman (Eds.), *Computers and Thought* (pp. 39–70). New York: McGraw-Hill.

Niantic. (2021). "Niantic Opens Lightship Platform Globally, Empowering Developers to Build Their Vision for the Real-World Metaverse." *Niantic Blog* (November 8, 2021). Accessed January 5, 2022. https://nianticlabs.com/blog/lightshiplaunch/?hl=en.

Nielsen, Jakob. (1989). "Trip Report: International Conference on Fifth Generation Computer Systems 1988 Tokyo, Japan, 28 November—2 December 1988." *SIGCHI Bulletin* 21, no. 1 (July 1989): 68–71.

Nielsen, Jakob. (1994). "Enhancing the Explanatory Power of Usability Heuristics." *CHI'94: Proceedings of the SIGCHI Conference on Human Factors in Computing Systems*, 152–158 (Boston, April 1994). New York: Association for Computing Machinery.

Nijholt, Anton. (2017). "Mischief Humor in Smart and Playable Cities." In Anton Nijholt (Ed.), *Playable Cities: The City as a Digital Playground* (pp. 235–253). Singapore: Springer.

Nilsson, Nils J. (2010). *The Quest for Artificial Intelligence: A History of Ideas and Achievements.* New York: Cambridge University Press.

Nitsche, Michael. (2008). *Video Game Spaces: Image, Play, and Structure in 3D Worlds.* Cambridge, MA: MIT Press.

Nohr, Rolf F. (2019). "The Development of *Decision Support Systems* in the 1960s as Antecedent of 'AI Rationality'." *Eludamos. Journal for Computer Game Culture* 10, no. 1: 67–90.

Nova, Nicolas. (2014). "The Technical Conditions of a Gameful World." In Steffen P. Walz and Sebastian Deterding (Eds.), *The Gameful World: Approaches, Issues, Applications* (pp. 395–404). Cambridge, MA: MIT Press.

NVIDIA. (2021). "NVIDIA Announces Platform for Creating AI Avatars." *NVIDIA Newsroom* (November 9, 2021). Accessed November 15, 2021. https://nvidianews.nvidia.com/news/nvidia-announces-platform-for-creating-ai-avatars?ncid=pa-so-twit-323872.

Oak Ridge National Laboratory (ORNL). (2021). "ORNL's Simulation Tool Creates Digital Twin of Buildings from Coast to Coast." *Oak Ridge National Laboratory News* (August 11, 2021). Accessed January 22, 2022. www.ornl.gov/news/ornls-simulation-tool-creates-digital-twin-buildings-coast-coast.

Ochs, Magalie, Nicolas Sabouret, and Vincent Corruble. (2009). "Simulation of the Dynamics of Nonplayer Characters' Emotions and Social Relations in Games." *IEEE Transactions on Computational Intelligence and AI in Games* 1, no. 4 (December 2009): 281–297.

Offenhuber, Dietmar. (2020). "What We Talk About When We Talk About Data Physicality." *IEEE Computer Graphics and Applications* 40, no. 6 (November/December 2020): 25–37.

Orkin, Jeff. (2006). "Three States and a Plan: The AI of F.E.A.R." *Game Developers Conference* (2006). Accessed November 30, 2021. www.gdcvault.com/play/1013282/Three-States-and-a-Plan and http://alumni.media.mit.edu/~jorkin/gdc2006_orkin_jeff_fear.pdf.

Osborn, Joseph, Noah Wardrip-Fruin, and Michael Mateas. (2017). "Refining Operational Logics." *FDG'17: Proceedings of the Twelfth International Conference on the Foundations of Digital Games* (August 2017), Article No. 27, 1–10. Accessed March 27, 2021. https://doi.org/10.1145/3102071.3102107.

Ouellette, Marc, and Steven Conway. (2019). "A Feel for the Game: AI, Computer Games and Perceiving Perception." *Eludamos. Journal for Computer Game Culture* 10, no. 1: 9–25.

Parker Brothers. (1978). "Merlin: The Electronic Wizard." Accessed January 5, 2022. www.theelectronicwizard.com/manual.pdf.

Peck, J. (2013). "Austere Reason, and the Eschatology of Neoliberalism's End Times." *Comparative European Politics* 11, no. 6 (November 2013): 713–721.

Peloton Interactive. (2020). "Peloton Investor and Analyst Session." *Peloton Events and Presentations* (September 15, 2020). Accessed January 5, 2022. https://investor.onepeloton.com/static-files/5155a9dc-1da8-4d6a-b232-3c231b8983b6.

Peloton Interactive, Inc. v. Flywheel Sports, Inc., 2:18-cv-00390, No. 199 (District of Delaware February 3, 2020). Accessed January 15, 2022. https://1g2apq2swm0928cj93zcx2n1-wpengine.netdna-ssl.com/wp-content/uploads/2020/02/FWSettlement.pdf.

Peña, Jorge et al. (2018). "Game Perspective-Taking Effects on Players' Behavioral Intention, Attitudes, Subjective Norms, and Self-Efficacy to Help Immigrants: The Case of *Papers, Please.*" *Cyberpsychology, Behavior, and Social Networking* 21, no. 11 (November 2018): 687–693.

Pérez-Latorre, Óliver. (2019). "Post-apocalyptic Games, Heroism and the Great Recession." *Game Studies: The International Journal of Computer Game Research* 19, no. 3 (December 2019). http://gamestudies.org/1903/articles/perezlatorre. Accessed August 27, 2020.

Piascik, Bob et al. (2010). *Technology Area 12: (Draft) Materials, Structures, Mechanical Systems, and Manufacturing Roadmap* (November 2010). Washington, DC: National Aeronautics and Space Administration. Accessed January 24, 2022. www.nasa.gov/pdf/501625main_TA12-MSMSM-DRAFT-Nov2010-A.pdf.

Pollack, Andrew. (1992). "'Fifth Generation' Became Japan's Lost Generation." *New York Times*, June 5, 1992, D1. Accessed August 7, 2021. www.nytimes.com/1992/06/05/business/fifth-generation-became-japan-s-lost-generation.html.

Prendinger, Helmut, and Mitsuru Ishizuka. (2001). "Social Role Awareness in Animated Agents." *AGENTS'01: Proceedings of the Fifth International Conference on Autonomous Agents*, 270–277 (Montreal, May 2001). New York: Association for Computing Machinery.

Preuss, Mike, and Sebastian Risi. (2020). "A Games Industry Perspective on Recent Game AI Developments: Interview with Duygu Cakmak, Creative Assembly." *Künstliche Intelligenz* 34, no. 1 (March 2020): 81–83.

Prug, Toni, and Paško Bilić. (2021). "Work Now, Profit Later: AI Between Capital, Labour and Regulation." In Phoebe V. Moore and Jamie Woodcock (Eds.), *Augmented Exploitation: Artificial Intelligence, Automation and Work* (pp. 30–40). London: Pluto Press.

Rankin, Joy Lisi. (2018). *A People's History of Computing in the United States*. Cambridge, MA: Harvard University Press.

Rasheed, Adil, Omer San, and Trond Kvamsdal. (2020). "Digital Twin: Values, Challenges and Enablers from a Modeling Perspective." *IEEE Access* 8 (January 28, 2020): 21980–22012. Accessed January 25, 2022. https://ieeexplore.ieee.org/stamp/stamp.jsp?tp=&arnumber=8972429.

Reed, Alison, and Amanda Phillips. (2013). "Additive Race: Colorblind Discourses of Realism in Performance Capture Technologies." *Digital Creativity* 24, no. 2: 130–144.

Research and Markets. (2021). *Home Fitness Equipment Global Market Report 2021: COVID-19 Implications and Growth to 2030.* September 2021 Report. Dublin, Ireland: Research and Markets.

Riegler, Gernot, and Vladlen Koltun. (2020). "Free View Synthesis." In Andrea Vedaldi, Horst Bischof, Thomas Brox, and Jan-Michael Frahm (Eds.), *Computer Vision—ECCV 2020*. Lecture Notes in Computer Science 12364 (pp. 623–640). Cham, Switzerland: Springer. Accessed October 30, 2021. https://arxiv.org/pdf/2008.05511.pdf.

Risi, Sebastian, and Mike Preuss. (2020). "From Chess and Atari to StarCraft and Beyond: How Game AI is Driving the World of AI." *Künstliche Intelligenz* 34, no. 1 (March 2020): 7–17.

Risi, Sebastian, and Julian Togelius. (2020). "Increasing Generality in Machine Learning Through Procedural Content Generation." *Nature Machine Intelligence* 2 (August 2020): 428–436.

Rodriguez, Hector. (2006). "The Playful and the Serious: An Approximation to Huizinga's *Homo Ludens*." *Game Studies: The International Journal of Computer Game Research* 6, no. 1 (December 2006). http://gamestudies.org/0601/articles/rodriges. Accessed August 27, 2020.

Roitman, Janet. (2014). *Anti-Crisis*. Durham, NC: Duke University Press.

Rubin, Courtney. (2021). "'The Netflix of Wellness': Inside the Hollywoodization of Peloton." *The Hollywood Reporter* (June 14, 2021). Accessed January 10, 2022. www.hollywoodreporter.com/business/business-news/hollywoodization-of-peloton-1234964386.

Russell, Benson. (2010). "The Secrets of Enemy AI in *Uncharted 2*." *Gamasutra* (November 3, 2010). Accessed April 24, 2020. www.gamasutra.com/view/feature/134566/the_secrets_of_enemy_ai_in_.php.

Salen, Katie, and Eric Zimmerman. (2004). *Rules of Play: Game Design Fundamentals*. Cambridge, MA: MIT Press.

Samuel, Arthur L. (1959). "Some Studies in Machine Learning Using the Game of Checkers." *IBM Journal of Research and Development* 3, no. 3 (July 1959): 210–229. Accessed August 1, 2021. https://ieeexplore.ieee.org/stamp/stamp.jsp?tp=&arnumber=5392560.

Santosh, K. C. (2020). "AI-Driven Tools for Coronavirus Outbreak: Need of Active Learning and Cross-Population Train/Test Models on Multitudinal/Multimodal Data." *Journal of Medical Systems* 44, no. 5 (May 2020): 1–5. Accessed September 5, 2021. www.ncbi.nlm.nih.gov/pmc/articles/PMC7087612/pdf/10916_2020_Article_1562.pdf.

Savage, Neil. (2022). "Virtual Duplicates: Digital Twins Aim to Model Reality So We Can See How It Changes." *Communications of the ACM* 65, no. 2 (February 2022): 14–16.

Schreier, Jason. (2017). *Blood, Sweat, and Pixels: The Triumphant, Turbulent Stories Behind How Video Games Are Made*. New York: Harper.

Schröter, Felix, and Jan-Noël Thon. (2014). "Video Game Characters: Theory and Analysis." *Diegesis* 3, no. 1: 40–77.

Shannon, Claude E. (1948). "A Mathematical Theory of Communication." *The Bell System Technical Journal* 27 (July, October 1948): 379–423, 623–656.

Shannon, Claude E. (1950a). "A Chess-Playing Machine." *Scientific American* 182, no. 2 (February 1950): 48–51.

Shannon, Claude E. (1950b). "Programming a Computer for Playing Chess." *Philosophical Magazine*, Seventh Series, 41, no. 314 (March 1950): 256–275.

Shen, Zhuoqian, and Suiping Zhou. (2006). "Behavior Representation and Simulation for Military Operations on Urbanized Terrain." *SIMULATION* 82, no. 9 (September 2006): 593–607.

Simon, Herbert A. (1991). "Artificial Intelligence: Where Has It Been, and Where Is It Going?" *IEEE Transactions on Knowledge and Data Engineering* 3, no. 2 (June 1991): 128–136.

Simon, Herbert A., and Allen Newell. (1958). "Heuristic Problem Solving: The Next Advance in Operations Research." *Operations Research* 6, no. 1 (January–February 1958): 1–10.

Sims, David. (2020). "*The Last of Us Part II* Tests the Limits of Video-Game Violence." *The Atlantic* (July 1, 2020). Accessed July 8, 2020. www.theatlantic.com/culture/archive/2020/07/the-last-of-us-limits-video-game-violence/613696/.

Sindreu, Jon. (2021). "Covid-19 Wrecked the Algorithms That Set Airfares, but They Won't Stay Dumb." *Wall Street Journal* (May 17, 2021). Accessed May 17, 2021. www.wsj.com/articles/covid-19-wrecked-the-algorithms-that-set-airfares-but-they-wont-stay-dumb-11621253268.

Singh, Satinder et al. (2010). "Intrinsically Motivated Reinforcement Learning: An Evolutionary Perspective." *IEEE Transactions on Autonomous Mental Development* 2, no. 2 (June 2010): 70–82.

Smart, Paul R. (2018). "Human-Extended Machine Cognition." *Cognitive Systems Research* 49 (June 2018): 9–23.

Smith, Gillian. (2015). "An Analog History of Procedural Content Generation." *FDG'15: Proceedings of the Tenth International Conference on the Foundations of Digital Games* (Pacific Grove, CA, June 2015). Accessed September 18, 2021. www.fdg2015.org/papers/fdg2015_paper_19.pdf.

Sony. (1998). "Sony Develops OPEN-R Architecture for Entertainment Robots." *Sony Information: News: Press Archive.* June 10, 1998. Accessed August 29, 2021. www.sony.com/en/SonyInfo/News/Press_Archive/199806/98-052/.

Sony. (1999). "Raising AIBO—The Handbook." Accessed July 17, 2021. www.sony-aibo.com/wp-content/uploads/2013/01/ERS111AIBO_Hbook.pdf.

Sony. (2002). "OPEN-R Architecture Specifications to Be Made Public." *Sony Information: News: Press Archive.* May 7, 2002. Accessed August 29, 2021. www.sony.com/en/SonyInfo/News/Press_Archive/200205/02-017E/.

Sony. (2021). "Creating Lifestyles Where Robots and Humans Make Emotional Connections: AIBO." *About Sony: Corporate Info: AI Initiatives.* Accessed July 17, 2021. www.sony.com/en/SonyInfo/sony_ai/aibo.html.

Sotamaa, Olli. (2014). "Artifact." In Mark J. P. Wolf and Bernard Perron (Eds.), *The Routledge Companion to Video Game Studies* (pp. 3–9). New York: Routledge.

Star, Susan Leigh. (1999). "The Ethnography of Infrastructure." *American Behavioral Scientist* 43, no. 3 (November 1999): 377–391.

Star, Susan Leigh, and Karen Ruhleder (1996). "Steps Toward an Ecology of Infrastructure: Design and Access for Large Information Spaces." *Information Systems Research* 7, no. 1 (March 1996): 111–134.

Steinberg, Marc. (2019). *The Platform Economy: How Japan Transformed the Consumer Internet.* Minneapolis: University of Minnesota Press.

Stenros, Jaakko. (2014). "Behind Games: Playful Mindsets and Transformative Practices." In Steffen P. Walz and Sebastian Deterding (Eds.), *The Gameful World: Approaches, Issues, Applications* (pp. 201–222). Cambridge, MA: MIT Press.

Straeubig, Michael. (2019). "Games, AI, and Systems." *Eludamos. Journal for Computer Game Culture* 10, no. 1: 141–160.

Strong, Christina R., and Michael Mateas. (2008). "Talking with NPCs: Towards Dynamic Generation of Discourse Structures." *Proceedings of the Fourth Artificial Intelligence and Interactive Digital Entertainment Conference,* 114–119. Menlo Park, CA: Association for the Advancement of Artificial Intelligence Press.

Sutherland, Gordon B. B. M. (1959). "The Mechanization of Thought Processes." *Proceedings of a Symposium on the Mechanisation of Thought Processes* (Volume 1), ix–x (National Physical Laboratory, Teddington, England, November 24–27, 1958). London: Her Majesty's Stationery Office.

Švelch, Jaroslav. (2020). "Should the Monster Play Fair? Reception of Artificial Intelligence in *Alien: Isolation*." *Game Studies: The International Journal of Computer Game Research* 20, no. 2 (June 2020). http://gamestudies.org/2002/articles/jaroslav_svelch. Accessed August 27, 2020.

Take-Two Interactive Software. (2020). "System and Method for Virtual Navigation in a Gaming Environment." *United States Patent Application Publication US 2020/0338450 A1* (October 29, 2020). Accessed September 18, 2021. https://patentimages.storage.googleapis.com/9d/f0/4e/caa93c4f8160e2/US20200338450A1.pdf.

Tassi, Eduardo. (2020). "Level Design Analysis: Oxenfurt Level in *The Witcher 3: Wild Hunt*." *Gamasutra* (August 12, 2020). Accessed August 27, 2020. https://gamasutra.com/blogs/EduardoTassi/20200812/367874/Level_Design_Analysis_Oxenfurt_level_in_The_Witcher_3_Wild_Hunt.php?elq_mid=99147&elq_cid=26826733.

Telemetrics. (2017). "Peloton." *Telemetrics Marketing Case Study* (March 30, 2017). Accessed January 10, 2022. www.telemetricsinc.com/casestudypeloton033017.

Thacker, Eugene. (2004). "Foreword: Protocol Is as Protocol Does." In Alexander R. Galloway (Ed.), *Protocol: How Control Exists after Decentralization* (pp. xi–xxii). Cambridge, MA: MIT Press.

Thompson, Tommy. (2014). "In the Director's Chair: The AI of *Left 4 Dead*." *Medium* (December 1, 2014). Accessed September 8, 2021. https://medium.com/@t2thompson/in-the-directors-chair-the-ai-of-left-4-dead-78f0d4fbf86a.

Thompson, Tommy. (2017). "The Perfect Organism: The AI of *Alien: Isolation*." *Game Developer* (October 31, 2017). Accessed September 8, 2021. www.gamedeveloper.com/design/the-perfect-organism-the-ai-of-alien-isolation.

Thompson, Tommy (2020a). "Endure and Survive: The AI of *The Last of Us*." *Game Developer* (June 17, 2020). Accessed September 8, 2021. www.gamedeveloper.com/design/endure-and-survive-the-ai-of-the-last-of-us.

Thompson, Tommy. (2020b). "Revisiting the AI of *Alien: Isolation*." *Game Developer* (May 20, 2020). Accessed September 8, 2021. www.gamedeveloper.com/design/revisiting-the-ai-of-alien-isolation.

Thon, Jan-Noël. (2019). "Playing with Fear: The Aesthetics of Horror in Recent Indie Games." *Eludamos. Journal for Computer Game Culture* 10, no. 1: 197–231.

Togelius, Julian. (2018). *Playing Smart: On Games, Intelligence, and Artificial Intelligence.* Cambridge, MA: MIT Press.

Tonal. (2018). "Digital Weights—The Future of Fitness. Today." *Tonal Blog: Product News* (August 16, 2018). Accessed January 5, 2022. www.tonal.com/blog/digital-weights-the-future-of-fitness-today.

Tozour, Paul. (2002a). "The Evolution of Game AI." In Steve Rabin (Ed.), *AI Game Programming Wisdom* (pp. 3–15). Hingham, MA: Charles River Media.

Tozour, Paul. (2002b). "First-Person Shooter AI Architecture." In Steve Rabin (Ed.), *AI Game Programming Wisdom* (pp. 387–396). Hingham, MA: Charles River Media.

Treanor, Mike et al. (2015). "AI-Based Game Design Patterns." *FDG'15: Proceedings of the Tenth International Conference on the Foundations of Digital Games* (Pacific Grove, CA, June 2015). Accessed September 18, 2021. www.fdg2015.org/papers/fdg2015_paper_23.pdf.

Tsigkari, Martha et al. (2021). "On-site with Spot: Robotic Refinement of Workflow Procedures." *Foster + Partners Plus* (December 20, 2021). Accessed February 15, 2022. www.fosterandpartners.com/plus/on-site-with-spot/.

Turing, Alan M. (1948). "Intelligent Machinery: A Report by A. M. Turing." National Physical Laboratory, London, UK. Accessed July 15, 2021. www.npl.co.uk/getattach ment/about-us/History/Famous-faces/Alan-Turing/80916595-Intelligent-Machinery. pdf?lang=en-GB.

Turing, Alan M. (1950). "Computing Machinery and Intelligence." *Mind: A Quarterly Review of Psychology and Philosophy* 59, no. 236 (October 1950): 433–460.

Turing, Alan M. (1996). "Intelligent Machinery, A Heretical Theory." *Philosophia Mathematica* 4, no. 3 (September 1996): 256–260.

Turkle, Sherry. (2021). *The Empathy Diaries: A Memoir.* New York: Penguin Press.

Vodopivec, Tom, Spyridon Samothrakis, and Branko Šter. (2017). "On Monte Carlo Tree Search and Reinforcement Learning." *Journal of Artificial Intelligence Research* 60: 881–936.

Voorhees, Gerald. (2014). "Play and Possibility in the Rhetoric of the War on Terror: The Structure of Agency in *Halo 2.*" *Game Studies: The International Journal of Computer Game Research* 14, no. 1 (August 2014). http://gamestudies.org/1401/articles/gvoorhees. Accessed August 27, 2020.

Wang, Hao, and Chuen-Tsai Sun. (2011). "Game Reward Systems: Gaming Experiences and Social Meanings." *DiGRA'11—Proceedings of the 2011 DiGRA International Conference: Think Design Play* 6 (September 2011). Accessed May 7, 2020. www.digra.org/wp-content/uploads/digital-library/11310.20247.pdf.

Wardrip-Fruin, Noah. (2005). "Playable Media and Textual Instruments." *Dichtung Digital* 5, no. 34. Accessed March 27, 2021. www.dichtung-digital.de/2005/1/Wardrip-Fruin/index.htm.

Wardrip-Fruin, Noah. (2009). *Expressive Processing: Digital Fictions, Computer Games, and Software Studies.* Cambridge, MA: MIT Press.

Wardrip-Fruin, Noah, Michael Mateas, Steven Dow, and Serdar Sali. (2009). "Agency Reconsidered." *DiGRA'09—Proceedings of the 2009 DiGRA International Conference: Breaking New Ground: Innovation in Games, Play, Practice and Theory* 5 (September 2009). Accessed March 27, 2021. www.digra.org/wp-content/uploads/digital-library/09287.41281.pdf.

Ware, Willis H. (2008). *RAND and the Information Evolution: A History in Essays and Vignettes.* Santa Monica, CA: RAND.

Wark, McKenzie. (2007). *Gamer Theory.* Cambridge, MA: Harvard University Press.

Warpefelt, Henrik, and Magnus Johansson. (2019). "Understanding the User Experience of AI Through the Lens of Game Studies." *Eludamos. Journal for Computer Game Culture* 10, no. 1: 183–195.

Watterson, Scott R. et al. (2001). "Systems and Methods for Providing an Improved Exercise Device with Motivational Programming." *United States Patent Number US 6,312,363 B1* (November 6, 2001). Accessed January 15, 2022. https://patentimages.storage.googleapis.com/da/e3/af/6dbb9b7f7b5145/US6312363.pdf

Watterson, Scott R. et al. (2020). "Coordinated Weight Selection." *United States Patent Number US 10,864,407 B2* (December 15, 2020). Accessed January 15, 2022. https://patentimages.storage.googleapis.com/41/80/fb/926f93ea5c3740/US10864407.pdf.

Weaver, Warren. (1949). "The Mathematics of Communication." *Scientific American* 181, no. 1 (July 1949): 11–15.

Weaver, Warren. (1961). "A Quarter Century in the Natural Sciences." *Public Health Reports* 76, no. 1 (January 1961): 57–65.

Weiser, Mark. (1991). "The Computer for the Twenty-First Century." *Scientific American* 265, no. 3 (September 1991): 94–104.

Weizenbaum, Joseph. (1976). *Computer Power and Human Reason: From Judgment to Calculation.* San Francisco: W. H. Freeman and Company.

Whitby, Blay. (1986). "The Computer as a Cultural Artefact." In Karamjit S. Gill (Ed.), *Artificial Intelligence for Society* (pp. 115–124). New York: John Wiley and Sons.

White, Topher. (2018). "The Fight Against Illegal Deforestation with TensorFlow." *Google: The Keyword* blog (March 21, 2018). Accessed March 29, 2021. www.blog.google/technology/ai/fight-against-illegal-deforestation-tensorflow/.

Whitson, Jennifer. R. (2013). "Gaming the Quantified Self." *Surveillance & Society* 11, no. 1/2: 163–176.

Whitson, Jennifer R. (2014). "Foucault's Fitbit: Governance and Gamification." In Steffen P. Walz and Sebastian Deterding (Eds.), *The Gameful World: Approaches, Issues, Applications* (pp. 339–358). Cambridge, MA: MIT Press.

Whitten, Marc. (2021). "Welcome, Weta Digital!" *Unity Blog: News* (November 9, 2021). Accessed January 5, 2022. https://blog.unity.com/news/welcome-weta-digital.

Wilson, Douglas. (2011). "Brutally Unfair Tactics Totally OK Now: On Self-Effacing Games and Unachievements." *Game Studies: The International Journal of Computer Game Research* 11, no. 1 (February 2011). http://gamestudies.org/1101/articles/wilson. Accessed August 27, 2020.

Winnicott, Donald Woods. (1971). *Playing and Reality.* London: Tavistock Publications.

Wolf, Gary. (2009). "Know Thyself: Tracking Every Facet of Life, from Sleep to Mood to Pain, 24/7/365." *Wired* (June 22, 2009). Accessed August 25, 2021. www.wired.com/2009/06/lbnp-knowthyself/.

Wolf, Gary. (2010). "The Data-Driven Life." *New York Times Magazine* (April 28, 2010). Accessed August 25, 2021. www.nytimes.com/2010/05/02/magazine/02self-measurement-t.html.

Woodcock, Jamie. (2021). *The Fight Against Platform Capitalism: An Inquiry into the Global Struggles of the Gig Economy.* London: University of Westminster Press.

Woodcock, Steven. (1998). "Game AI: The State of the Industry." *Gamasutra* (November 20, 1998). Accessed March 28, 2021. www.gamasutra.com/view/feature/131705/game_ai_the_state_of_the_industry.php.

Wordsworth, Saul. (2019). "Double Take: What Role Do Digital Twins Play in Helping Airports Improve Operations Now and in the Future?" *Passenger Terminal World* (September 2019): 28–34. Accessed January 23, 2022. www.ukimediaevents.com/publication/a474db3d/30.

Yannakakis, Georgios N., and Julian Togelius. (2015). "A Panorama of Artificial and Computational Intelligence in Games." *IEEE Transactions on Computational Intelligence and AI in Games* 7, no. 4: 317–335.

Yannakakis, Georgios N., and Julian Togelius. (2018). *Artificial Intelligence and Games.* Cham, Switzerland: Springer.

Yannakakis, Georgios N. et al. (2013). "Player Modeling." In Simon Lucas, Michael Mateas, Mike Preuss, Pieter Spronck, and Julian Togelius (Eds.), *Artificial and Computational Intelligence in Games: A Follow-up to Dagstuhl Seminar 12191* (pp. 45–59). Saarbrücken/Wadern, Germany: Dagstuhl Publishing.

Yoshida, Kenichiro. (2021). *Sony: Corporate Strategy Meeting* (May 26, 2021): 1–61. Accessed September 18, 2021. www.sony.com/en/SonyInfo/IR/library/presen/strategy/pdf/2021/speech_E.pdf.

Zimmerman, Eric. (2009). "Gaming Literacy: Game Design as a Model for Literacy in the Twenty-First Century." In Bernard Perron and Mark J. P. Wolf (Eds.), *The Video Game Theory Reader 2* (pp. 23–31). New York: Routledge.

Zylinska, Joanna. (2020). *Perception at the End of the World (or How Not to Play Video Games)*. Pittsburgh: Flugschriften. Accessed May 4, 2021. https://flugschriftencom.files.wordpress.com/2020/04/flugschriften-7-joanna-zylinska-perceptions-at-the-end-of-the-world-or-how-not-to-play-videogames.pdf.

INDEX